US Climate
Variability &
Predictability
Program
Science Plan

US CLIVAR Science Plan

US CLIVAR Scientific Steering Committee

December 2013

The US CLIVAR Science Plan was prepared with federal support of NASA (AGS-0963735), NOAA (NA11OAR4310213), NSF (AGS-0961146), and DOE (AGS-1357212) by the US CLIVAR Project Office as administered by the University Corporation for Atmospheric Research, an independent research organization that is not an agency of the US Government. Any opinions, findings and conclusions, or recommendations expressed herein are those of the author(s) and do not necessarily reflect the views of the sponsoring funding agencies.

Bibliographic citation:
US CLIVAR Scientific Steering Committee, 2013: US Climate Variability & Predictability Program Science Plan. Report 2013-7, US CLIVAR Project Office, Washington, DC 20005.

Table of Contents

List of Figures and Tables

Executive Summary

The mission of the US Climate Variability & Predictability (CLIVAR) Program is: **To foster understanding and prediction of climate variability and change on intraseasonal-to-centennial timescales, through observations and modeling with emphasis on the role of the ocean and its interaction with other elements of the Earth system, and to serve the climate community and society through the coordination and facilitation of research on outstanding climate questions.**

Since its inception, US CLIVAR-led research has played a substantial role in advancing our understanding of, and skill in predicting, climate variability and change. Advances include:

- Significant increases in understanding of the climate system and its predictability;
- Expansion of a sustained ocean observing system;
- Development and coordination of inter-comparisons of ocean and coupled simulations that have led to improved predictive capability;
- Development of climate models with improved representation of physical processes;
- Integrated Earth-system science and modeling that broadens the interdisciplinary perspective of climate science;
- Regular assessments of the changing climate system, together with its impacts on human and natural systems to establish a sound scientific basis for developing mitigation and adaptation options; and
- Increased attention to the uncertainties and confidence limits of both observed and predicted climate information.

These advances have been motivated by fundamental science questions, which also guide and drive US CLIVAR activities. These fundamental science questions include:

- What processes are critical for determining climate variability and change related to the ocean?
- What are the connections and feedbacks between oceanic climate variability and other components of the Earth's climate system?
- How predictable is the climate on different time and space scales?
- What determines regional expressions of climate variability and change?

Pursuit of these questions led to the focus of US CLIVAR since its inception, and they remain just as important today. They cover a range of climatic issues: from basic understanding of climate processes to what aspects of the climate system can be predicted on global-to-local scales.

The solid progress made over the last 15 years calls for a review and an update of the original terms of reference for US CLIVAR. This Science Plan updates the goals and priorities of US CLIVAR in light of the achievements to date. Additionally, the Science Plan articulates important implementation activities to expand upon US CLIVAR's core research to target specific Research Challenges that emphasize strengthened ties to the broader Earth science community and relevance to societal impacts. As such, the Science Plan provides a guidebook for the maintenance and development of scientific activities during the lifetime of the program.

To achieve its Mission, US CLIVAR has the following goals:

- Understand the role of the oceans in observed climate variability on different timescales.
- Understand the processes that contribute to climate variability and change in the past, present, and future.
- Better quantify uncertainty in the observations, simulations, predictions, and projections of climate variability and change.
- Improve the development and evaluation of climate simulations and predictions.
- Collaborate with research and operational communities that develop and use climate information.

The future US CLIVAR will continue the research agenda of the original program. At the same time, it will target specific Research Challenges involving the observational, modeling, and prediction communities of US CLIVAR. US CLIVAR highlighted these Research Challenges as topical themes in recent years. Given their complex cross-disciplinary nature, progress in these areas can benefit from US CLIVAR facilitation. Each has its own set of defining questions and science issues that are directly related to US CLIVAR's overarching goals. Four Research Challenges are currently identified:

- Decadal variability and predictability
- Climate extremes
- Polar climate
- Climate and ocean carbon/biogeochemistry

It is believed that progress and coordination in these areas will benefit many of the core research interests of US CLIVAR. These four Challenges are expected to remain as focus areas for the next decade and beyond, but will be reviewed periodically to determine continuance. Additional Challenges will be considered and taken on as US CLIVAR makes progress on this initial set.

Progress on all US CLIVAR goals requires proper strategies and tools to identify and simulate leading-order climate processes and thereby to enable predictions that are as accurate and reliable as possible. For example, the creation and maintenance of adequate observational networks are essential for all US CLIVAR activities. A key strategy is therefore to assess the adequacy of historical data records and the existing ocean observing system, to sustain and evolve critical observing capabilities,

and to determine the additional observations that are needed to foster understanding of climate processes and variability. In addition, advanced models, assimilations, prediction, and verification techniques need to be established to enable predictions of climate variability and change with quantified uncertainty. Such strategies are cross-cutting for most, if not all, of US CLIVAR's research activities, and they comprise a way forward for advancing US CLIVAR's science goals.

US CLIVAR management structure facilitates close collaboration between the climate science community and the funding agencies that sponsor climate research. It consists of: a Scientific Steering Committee (SSC) and three panels comprised of research community members; an Inter-Agency Group (IAG) of the program managers who fund US CLIVAR research and planning efforts; and a Project Office. US CLIVAR activities will continue to include those that have proven successful, including: Climate Process Teams, Working Groups, and Science Teams; support for meetings and workshops; providing opportunities for young investigators; and facilitation of agency solicitations and project awards.

US CLIVAR progress to date has increased public awareness of the impacts of climate variability on the safety and well-being of society. It has accomplished this through improved observing, understanding, modeling, and predicting our climate system. Continued advancement is needed in all these areas. As appreciation evolves of the interconnectedness of the physical, biological, and chemical elements of the Earth system that impact on and are impacted by climate, US CLIVAR is compelled to engage with other Earth science communities to improve our understanding of the oceans' role in climate variability and change and provide relevant information for the future. CLIVAR's strength in facilitation of coordinated science within the United States and with the international climate community provides a strong leadership position. To achieve its goals, US CLIVAR will actively engage with other Earth science communities, often at the interface of traditional disciplinary boundaries. It will also provide support for the infrastructure needed for climate research, including observing systems, data centers, research platforms, modeling and prediction centers, and national and international scientific assessments.

US CLIVAR was established to facilitate investigation of the variability and predictability of the global climate system on intraseasonal-to-centennial timescales, with emphasis on the role of the ocean. US CLIVAR provides coordination for climate scientists involved in a wide-range of activities to advance our understanding and predictive ability of the Earth's climate. As part of that broader engagement, US CLIVAR also provides a US government interagency mechanism for coordinated US engagements in International CLIVAR.

Chapter 1

Introduction

C limate affects everyone. The change of seasons influences cultural traditions, agricultural practices, and even disease transmission. Year-to-year changes in climate, and the associated changes in weather characteristics, can either amplify or weaken the seasonally varying climate fluctuations, leading to floods or failed rainy seasons, cold spells or heat waves, all of which have substantial human, ecological, and economic consequences. Multi-year expressions of climate variability can yield periods of particularly devastating impacts, such as the Dust Bowl of the 1930s with its prolonged drought and high temperatures that crippled agriculture in the central United States and led to widespread migration out of that region. Conversely, prolonged periods of anomalous climate can be beneficial, such as enhanced rainfall in semi-arid regions that enable expansion of agriculture or drier than normal conditions that suppress vector-borne diseases like malaria. Along with natural variability, man-made climate change is part of the climate that we experience from year-to-year, decade-to-decade, and longer. Decades of observational and modeling research have shown the oceans to be a critical player in influencing the Earth's climate across timescales. In terms of our current ability to predict intraseasonal-to-decadal variability, particularly over the United States, the oceans are the dominant factor.

With the understanding that climate variability and change are global in scope, the World Climate Research Program (WCRP) was established in 1980 to address climate issues world-wide, under the joint sponsorship of the International Council for Science (ICSU) and the World Meteorological Organization (WMO) and, since 1993, also by the Intergovernmental Oceanographic Commission (IOC of UNESCO). The WCRP consists of four projects, each of which focuses on a key component of the Earth's climate system: CLIVAR (ocean-atmosphere), GEWEX (land-atmosphere), CliC (cryosphere), and SPARC (stratosphere). The main objectives, set for the WCRP are to determine the predictability of, and the effects of human activities on, climate.

The US Climate Variability & Predictability (CLIVAR) Program was established in 1997 as a focused United States contribution to the international CLIVAR project of the WCRP. It is supported by US science funding agencies: National Oceanic and Atmospheric Administration (NOAA), National Science Foundation (NSF), National Aeronautics and Space Administration (NASA), Department of Energy (DoE), and the Office of Naval Research (ONR). The activities of US CLIVAR are slightly broader than international CLIVAR, in that the other components of the climate system are included. However, its focus remains on the ocean's role in Earth's climate variability and change; consideration of land, cryosphere, or stratosphere in US CLIVAR is primarily through their interaction with seasonal-to-centennial changes in the ocean.

1.1 Importance of the ocean to climate

Oceans cover roughly three-quarters of our planet. Because water has a larger capacity for storing heat than air or land, ocean temperatures change more slowly. Changes and anomalies in the patterns of ocean surface temperatures modify where heat is then exchanged between the ocean and atmosphere. This can affect atmospheric pressure and thus the winds, which then influence ocean currents and further impact ocean-atmosphere heat exchange, thereby feeding back onto the patterns of ocean surface temperatures. The oceans also provide the primary source of moisture to the atmosphere, and so influence large-scale rainfall and air-temperature patterns, as well as the distribution of clouds and ice. Because clouds and ice mediate the amount of the sun's energy that is absorbed at the surface of the Earth, they also participate in feedback processes that influence ocean temperatures, which in turn influence regional and global climate.

Interactions between the ocean and atmosphere occur over many time and space scales and ultimately lead to large-scale changes in the atmospheric circulation and the climate that we experience. Weekly-to-monthly variability in regional weather and climate, often associated with extremes such as hurricanes and flooding, arise from air-sea interactions such as the Madden Julian Oscillation. Year-to-year changes in tropical ocean temperature, such as those associated with the El Niño-Southern Oscillation (ENSO) phenomenon that warms the eastern equatorial Pacific, play a major role in seasonal climate variability. About 20-30% of the land surface experiences statistically robust rainfall anomalies during El Niño and La Niña events. Over the past 30 years, we have greatly increased our observations, simulations, and predictive capability of ENSO. Subsequently, through a better understanding of the physical processes involved in its development and evolution, ENSO has arisen as the leading source of skill in seasonal forecasts. Nevertheless, fundamental questions concerning ocean-atmosphere connections and feedbacks still remain, as indicated by the fact that advances in coupled models continue to provide only incremental improvements in ENSO forecasts and by the presence of large biases in model sea surface temperatures (SSTs) in the eastern boundary regions. Decade-to-decade changes in ocean surface temperatures also appear to be an important driver of large-scale pluvial and drought periods over the United States and other parts of the world, as well as changes in Atlantic hurricane activity, but our understanding and ability to simulate oceanic decadal variability is still at an embryonic stage. The oceans also play a critical, though not entirely understood, role in Earth's response to anthropogenic climate change. The ability of the ocean to store tremendous amounts of heat and CO_2 has mitigated some of the global warming that we may otherwise have experienced. However, it seems the oceans do not do this uniformly in space or time, and the ability of the ocean to continue to absorb CO_2 from the atmosphere at present rates is not likely to continue. The absorption of heat at high latitudes is considered to be a primary cause of dramatic losses in sea ice, which may be partly responsible for recent increases in mid-latitude extreme weather and climate events. The absence of reflective ice will certainly be responsible for enhanced global warming in the future. At the same time, heating of the ocean leads to thermal expansion and a significant contribution to sea level rise.

Understanding the relationship between the timescales of variability in the ocean and other components of the climate system is essential for improving our ability to develop and use climate information. At the same time we need to build understanding of the spatial scales of the modes of variability and how variability in a given region can be influenced not only by local phenomena but also by modes with larger scales, such as ENSO, stemming from oceanic conditions far from that region. This understanding is also essential for building climate literacy in the public and governmental domains, which is necessary in order for decision-makers to develop well-informed policies.

1.2 Need for US CLIVAR

US CLIVAR was established to investigate the variability and predictability of the global climate system on intraseasonal-to-centennial timescales, with emphasis on the role of the ocean. US CLIVAR provides coordination for climate scientists involved in a wide-range of activities to advance our understanding and predictive ability of the Earth's climate. Scientists are engaged in several capacities, both explicitly in US CLIVAR panels and working groups and implicitly through contributions to many of the opportunities funded by US CLIVAR's funding partners. For example, some scientists may be engaged in a planning capacity to identify promising avenues of research and perceived research gaps. Other scientists may be engaged in targeted working groups to make progress on specific and timely topics. US CLIVAR scientists also engage with the broader international community and across disciplines in Earth science to more efficiently build on community-wide efforts such as fostering sustained, effective observational networks, and developing physical models that can be used for hypothesis testing as well as reliable predictions. As part of that broader engagement, US CLIVAR provides a US government interagency mechanism for coordinated US engagements in International CLIVAR.

Since its inception, US CLIVAR-led research has played a substantial role in advancing our understanding of and skill in predicting climate variability and change. Advances include:

- Significant increases in understanding of the climate system and its predictability;
- Expansion of a sustained ocean observing system;
- Development and coordination of inter-comparisons of ocean and coupled simulations that have let to improved predictive capability;
- Development of climate models with improved representation of physical processes;
- Integrated Earth-system science and modeling that broadens the interdisciplinary perspective of climate science;
- Regular assessments of the changing climate system, its impacts on human and natural systems, and mitigation and adaptation options;
- Increased attention to the uncertainties and confidence limits of both observed and predicted climate information.

US CLIVAR contributions to these advances are reviewed in **Chapter 2** and the accompanying US CLIVAR Accomplishments Report.

Fundamental science questions that have guided and driven the advancement of US CLIVAR activities, and will continue to do so, include:

- What processes are critical for determining climate variability and change related to the ocean?
- What are the connections and feedbacks between oceanic climate variability and other components of the Earth's climate system?
- How predictable is the climate on different time and space scales?
- What determines regional expressions of climate variability and change?

These questions cover a range of climatic issues: from basic understanding of climate processes to what aspects of the climate system can be predicted on global-to-local scales. They are discussed in detail in **Chapter 3**.

US CLIVAR progress over the past 15 years has motivated, and

was motivated by, increased public awareness of the impacts of climate variability on the safety and well-being of society. In turn, this awareness has resulted in a shifting emphasis of national and international research, policy, and funding priorities. One notable outcome is operational climate services, the development of which has accelerated over the past 15 years with the advent of real-time seasonal climate forecasting. Climate services seek to address the increasing demand for climate information products by policy-makers and users in a wide range of sectors, from education to resource management to infrastructure planning. The development of such services requires continued evolution of the science that supports them. Continued advancement is needed in the quality of climate information, and an active role of the scientific research community is in observing, understanding, modeling, and predicting our climate system. It is incumbent on the climate community to convey the relevance of climate data and research to other academic and practitioner communities so that this research effort can provide sufficient benefits to society.

1.3 US CLIVAR Science Plan

The solid progress made over the last 15 years calls for an update of the original terms of reference for US CLIVAR. This Science Plan updates the goals and priorities of US CLIVAR based on the achievements to date. Additionally, the Science Plan articulates important implementation activities, including expanding upon US CLIVAR's core research to target specific Research Challenges (listed below) that emphasize strengthened ties to the broader Earth science community and relevance to societal impacts. As such, the Science Plan provides a guidebook for the maintenance and development of scientific activities during the lifetime of the program.

US CLIVAR's new 15-year science plan represents the interests of scientists and stakeholders throughout the climate community. Input was sought from the broader climate research community and funding agencies, including: funding agency interests; changing international and national program priorities; progress achieved over the past 15 years toward stated goals/objectives; and priority research topics and science questions framing the future program. Writing teams, consisting of scientists both inside US CLIVAR (current and former SSC members, panelists, Project Office personnel) and outside the program (experts on topics at the interface of disciplines) prepared an initial draft of the document. After a community-wide review, the SSC carefully considered all the review comments in revising the Science Plan for publication.

The next phase of US CLIVAR is envisioned to last 15 years. The first 15 years have built a strong foundation of improved understanding. At the same time, the community, national agencies and programs, and intergovernmental agencies and programs have made considerable progress toward building a global ocean observing system or GOOS. Elements of this observing system include global deployment of surface drifters and profiling Argo floats, arrays of moorings in the tropics, full depth sampling of the water column from ships of the repeat hydrography program, and moorings collecting time series at key extratropical locations. In some cases, more than a decade of observations is in hand that we did not have 15 years ago. Together with the sustained operation of GOOS going forward and with the progress made in improving models and computing facilities, we now in CLIVAR have a unique, new capability to investigate the variability and dynamics of the ocean, resolving many of the timescales, and capturing data from a time of change. Planning a new 15-year effort building on the past and fully utilizing the new datasets and new observing and modeling capabilities is a center piece of planning US CLIVAR.

In recognition of the long timescales and complexity of the mean climate and its variability, we believe meaningful progress in some areas can only be made over a significant period of time, allowing, for example, decadal modes to be observed. This planned 15-year timeframe is short enough to ensure that the goals of the new Science Plan remain relevant throughout the life of the program. At the same time, it is long enough to address the complex and deliberate processes of strengthening observing systems, incorporating knowledge gained from new observations into models, applying new observations and improved models towards simulation and prediction, and also interacting with scientists in other Earth components.

Mission and goals

At the heart of the new Science Plan is US CLIVAR's Mission:

To foster understanding and prediction of climate variability and change on intraseasonal-to-centennial timescales, through observations and modeling with emphasis on the role of the ocean and its interaction with other elements of the Earth system, and to serve the climate community and society through the coordination and facilitation of research on outstanding climate questions.

To achieve its Mission, US CLIVAR has the following goals:

- Understand the role of the oceans in observed climate variability on different timescales.
- Understand the processes that contribute to climate variability and change in the past, present, and future.
- Better quantify uncertainty in the observations, simulations, predictions, and projections of climate variability and change.
- Improve the development and evaluation of climate simulations and predictions.
- Collaborate with research and operational communities that develop and use climate information.

The goals are further elaborated in **Chapter 4**.

Research challenges

The new US CLIVAR will continue the research agenda of the original program, as outlined above and in Chapter 2. At the same time, it will target specific Research Challenges, which provide focused topics involving the observational, modeling, and prediction communities of US CLIVAR. Four Research Challenges are currently identified:

- Decadal variability and predictability
- Climate extremes
- Polar climate
- Climate and ocean carbon/biogeochemistry

US CLIVAR highlighted these Research Challenges as topical themes in recent years. Given their complex cross-disciplinary nature, progress in these areas can benefit from US CLIVAR facilitation. Each has its own set of defining questions and science issues, which are directly related to US CLIVAR's overarching goals, and they are discussed in **Chapter 5**. It is believed that progress and coordination in these areas will benefit many of the core research interests of US CLIVAR. These four Challenges are expected to remain as focus areas for the next decade and beyond, but they will be reviewed periodically to determine continuance and to initiate new ones. Additional Challenges will be considered and taken on as US CLIVAR makes progress on this initial set.

Cross-cutting strategies

Progress on all US CLIVAR activities requires proper strategies and tools to identify and simulate leading-order climate processes, and so to enable predictions that are as accurate and reliable as possible. For example, the creation and maintenance of adequate observational networks are essential for all US CLIVAR activities. A key strategy is therefore to assess the adequacy of historical data records and the existing ocean observing system, to sustain and evolve critical observing capabilities, and to determine the additional observations that are needed to foster understanding of climate processes and variability. In addition, advanced models, assimilations, prediction, and verification techniques need to be established to enable predictions of climate variability and change with quantified uncertainty. Such strategies are cross-cutting for most, if not all, of US CLIVAR's research activities, and they comprise a way forward for advancing US CLIVAR's science goals. They are outlined in **Chapter 6**.

Management and implementation

US CLIVAR management is designed to facilitate close collaboration between the climate science community and the funding agencies that sponsor climate research. It consists of: a Scientific Steering Committee (SSC) and three panels comprised of research community members; an Inter-Agency Group (IAG) of the program managers who fund US CLIVAR research and planning efforts; and a Project Office. US CLIVAR activities will continue to include the activities that have proven successful in the previous program: development of Climate Process Teams, Working Groups, and Science Teams; support for meetings and workshops; providing opportunities for young investigators; and facilitating agency solicitations and project awards. Most of US CLIVAR activities involve groups of experts in particular fields of climate research, so that typically US CLIVAR involves the participation of 100–200 volunteer scientists at any given time. US CLIVAR management and implementation activities are discussed in detail in **Chapter 7**.

Relationship to other programs

To achieve its goals, US CLIVAR will actively engage with other Earth science communities, often at the interface of traditional disciplinary boundaries. It will also provide support for the infrastructure needed for climate research, including observing systems, data centers, research platforms, modeling and prediction centers, and national and international scientific assessments. **Chapter 8** describes these envisioned collaborations.

Chapter 2

History and Achievements of US CLIVAR

An overview of the planning, evolution, and achievements of US CLIVAR during its first 15 years provides the background information for the development of the Fundamental Science Questions (Chapter 3) and Goals (Chapter 4) for the next era of US CLIVAR. It also serves to illustrate the types of activities that US CLIVAR will continue to pursue in addition to its new Research Challenges (Chapter 5).

2.1 History and key achievements

International origins

CLIVAR emerged internationally in 1995 with a science plan issued by the Joint Scientific Committee of the World Climate Research Program (WCRP 1995), emphasizing research to understand the variability of the climate system and its predictability arising from coupled ocean-atmosphere interactions. The new program built upon the successes of two WCRP programs focusing on the role of the ocean in climate, the Tropical Ocean Global Atmosphere (TOGA) and the World Ocean Circulation Experiment (WOCE), which were then coming to completion. The science plan organized research into a number of principal areas focused on variability and predictability regionally by ocean basins and regional monsoon systems, and globally for the study of anthropogenic climate change.

US program formulation

CS CLIVAR was launched in 1997 by an interagency group of science program managers at NASA, NOAA, NSF, and DoE seeking to coordinate the US contribution to International CLIVAR. The group appointed a Scientific Steering Committee (SSC) to identify science goals and guide implementation of the US program. Taking into account earlier science planning by the

National Research Council for separate programs on the Global Ocean-Atmosphere-Land System (NRC 1994; 1998a) and Decade-to-Century-Scale Climate Variability and Change (NRC 1995; 1998b), the SSC developed science goals for a single US program that aligned with the goals of International CLIVAR above. Specifically, the US CLIVAR goals were to:

- Identify and understand the major patterns of climate variability on seasonal and longer timescales and evaluate their predictability;
- Expand our capacity to predict short-term (seasonal to interannual) climate variability and search for ways to predict decadal variability;
- Better document the record of rapid climate changes in the past, as well as the mechanisms for these events, and evaluate the potential for abrupt climate changes in the future;
- Evaluate and enhance the reliability of models used to project climate change resulting from human activity, including anthropogenic changes in atmospheric composition; and
- Detect and describe any global climate changes that may occur.

Collectively, these goals promoted analysis and prediction of the state and evolution of climate spanning timescales from seasonal-to-centennial and spatial scales from regional-to-global.

Initial implementation

The SSC established three regional panels: the Atlantic, Pacific, and Pan-America, corresponding to International CLIVAR regions that most closely aligned with US regional interests and for which US planning was already mature. The three panels developed implementation plans identifying the initial research foci and their implementation needs for the Atlantic Climate Variability Experiment (Joyce and Marshall 2000), the Pacific Basin-wide Extended Climate Study (Davis et al. 2000), and US CLIVAR Pan American Research (Esbensen et al. 2002). Drawing upon the regional planning, the SSC issued the US CLIVAR Implementation Plan (US CLIVAR SSC 2000), outlining specific requirements and recommendations for US contributions to global and regional research. During the next five years, the three regional panels, in coordination with their International CLIVAR counterpart panels, organized observational enhancements, process studies and field campaigns, and modeling and predictability studies to explore climate variability in their respective regions. In addition to the three regional panels, US CLIVAR supported coordination and planning of several Working Groups and panels in International CLIVAR.

Mid-term reorganization

In 2005, US CLIVAR reorganized its panels to have a more global perspective, recognizing that work on regional coordination would continue to benefit from the regional panels of International CLIVAR. This decision was based on the fact that major climate signals are global in extent, making it limiting to study climate only in three individual regions. Because similar types of scientific activities are undertaken within each region, US CLIVAR reorganized its regional panels to reflect these activities, the new panels addressing Phenomena, Observations and Synthesis (POS), Process Studies and Model Improvement (PSMI), and Predictability, Prediction and Applications Interface (PPAI). This alternate structure was deemed advantageous, not only for expanding the geographic scope (e.g., to the Indian and Southern Oceans), but also for accelerating progress toward scientific deliverables including improved observing systems, climate models, analyses, and predictions. The SSC and these newly formed panels now meet jointly at annual US CLIVAR Summits.

Key achievements

Through its first 15 years, US CLIVAR has engaged the climate science community to plan and undertake coordinated science activities designed to observe, simulate, understand, and predict the global climate system and its impacts. Some of the program's most compelling and enduring achievements are highlighted here. Details on the motivation, activities, and impacts of each are provided in the accompanying US CLIVAR Accomplishments Report.

Exploration of climate modes and their relationship to the ocean

Modes (preferred patterns) of climate variability and their connections with the ocean have been of abiding interest to the climate community. US CLIVAR has made meaningful contributions towards describing these modes, understanding the underlying mechanisms, and diagnosing their impacts on natural and human systems.

Promotion of sustained and expanded in-situ observing systems

US CLIVAR has played a central role in motivating and establishing new and continuing in-situ ocean and atmosphere observing systems to monitor, understand and model climate variability, including the expansion of the Global Tropical Moored Buoy Array and establishment of Ocean Reference Stations, implementation of the global array of Argo profiling floats, building of an Atlantic Meridional Overturning Circulation (AMOC) observing system, and sustaining the global XBT network and the Global Drifter Program.

Evaluation and use of satellite products

Satellite data, providing a global view of the Earth's climate system, are crucial for US CLIVAR studies of climate variability and change and for improving the descriptive and predictive skill of general circulation models. US CLIVAR has helped establish the scientific justification and requirements for satellite missions, increasing awareness and community support for missions, and providing guidance on observational and scientific activities that should be considered in advance of and during the missions to improve the measurement, analysis, and utilization of remote sensing information.

Coordinated development of reanalyses

In-situ and satellite observing systems have provided strong impetus for the development of global ocean reanalysis. US CLIVAR and the Global Ocean Data Assimilation Experiment (GODAE) Program have collaborated to further accelerate the development of reanalysis systems and their utility. Recognizing the potential of coupled ocean-atmosphere assimilation in improving climate research and seasonal-interannual forecasts, US CLIVAR has also been the major driving force for developing a strategy for Integrated Earth System Analysis (IESA) to facilitate the study of the interaction among different components of the Earth's Climate System.

Support of US participation in process studies

The understanding of atmospheric and oceanic processes gained through process studies is vital for the development and improvement of empirical and dynamical models in their representation of the physical drivers of the climate system as well as for determining observational requirements for monitoring the evolution and prediction of the future state of the climate system. Since the start of US CLIVAR, US funding agencies have sponsored ten CLIVAR process studies to improve the understanding of coupled ocean-atmosphere processes in the eastern tropical Pacific, western boundary currents and mode waters in the North Pacific and Atlantic, mixing processes in the Southern Ocean, initiation of the Madden Julian Oscillation in the Indian Ocean, variability of upper-ocean salinity in the North Atlantic, and the monsoon systems of North and South America.

Best practices for process studies

Based on an evaluation of the success of these process studies in meeting their objectives, the US CLIVAR established a set of best practices to guide future community planning and implementation of process studies. The practices promote interaction of observationalists and modelers in planning and execution of data collection, analysis, and modeling synthesis activities in order to facilitate model improvements. They also advocate open, centralized data access and user-friendly data formatting to help ensure immediate use of data by scientists engaged in the process study and enable the long-term use of the data by the broader climate science community.

Establishment of Climate Process Teams

US CLIVAR has pioneered and implemented the concept of Climate Process Teams (CPTs) to facilitate and accelerate the transfer of climate process understanding gained through process studies and field campaigns to the development and testing of physical process parameterizations used in climate models. Each CPT assembles observation-oriented experimentalists, process modelers, process diagnosticians, and climate model developers into a single project focusing on a specific process that is poorly represented in components of climate models. US CLIVAR agencies have sponsored seven CPT projects to improve the fidelity of coupled climate models by improving the physical process representation in the component ocean and atmosphere GCMs, identify process studies needed to further refine models, and develop sustained observational requirements for climate model systems.

Sponsored diagnostics of late 19th-20th century simulations and 21st century projections

US CLIVAR implemented the Climate Model Evaluation Project (CMEP) to increase community-wide diagnostic research into the quality of Coupled Model Intercomparison Project (CMIP) simulations. CMEP promoted analyses of late 19th-20th century CMIP simulations through multi-model intercomparisons, ensembles, and comparisons with observational datasets. These analyses enable subsequent evaluations of the quality of global and regional climate projections of the 21st century in response to a range of future climate forcings. US CLIVAR funding agencies have sponsored 46 CMEP projects examining physical climate features such as oceanic and atmospheric modes of variability, regional climate and monsoon variability and trends, hydrological cycle behavior, and extreme events and coupled feedbacks among the ocean, atmosphere, sea ice, and carbon cycle.

Increased understanding the physical mechanisms of drought

US CLIVAR took several steps in recent years to advance both our understanding and ability to simulate and predict drought. The program sponsored a Drought Working Group, which proposed a working definition of drought, coordinated evaluations of existing relevant model simulations, coordinated new experiments to address outstanding uncertainties in the nature of drought, and coordinated and encouraged the analysis of observational datasets. The Drought In Coupled Models Project (DRICOMP) facilitated a community-wide analysis of simulations made with coupled atmosphere and ocean global circulation models to address issues such as the roles of the oceans and the seasonal cycle in drought, the impacts of drought on water availability, and distinctions between drought and drying. The DRICOMP and Working Group projects were complementary, providing a controlled assessment of the impact of past SST variations on drought, as well as assessing future drought under global warming scenarios.

Coordination of research to characterize predictability and improve predictions

US CLIVAR has facilitated several research efforts through limited lifetime working groups to investigate prediction skill, model fidelity, the dominant processes related to predictability, and the observed and simulated variations in dominant processes. The MJO Working Group developed a set of diagnostics to assess MJO simulation fidelity and forecast skill, and coordinated experiments to better understand and improve model representations and forecasts of the MJO. The Decadal Predictability Working Group investigated the predictability that may be realized from decadal-scale variability in addition to anthropogenic climate change. The group provided an overview of methodologies to separate decadal variations and anthropogenically forced trends, and they developed a framework to assess current capabilities for decadal predictions. More recently established working groups are evaluating predictability of tropical cyclone and hurricane activity under a warming climate (Hurricane WG), evaluating whether current climate models produce extremes for the right reasons and whether they can be used for predicting and projecting short-term extremes in temperature and precipitation over North America (Extremes WG); and clarifying, coordinating and synthesizing research toward a better understanding the diversity of ENSO, including surface and sub-surface characteristics, tropical-extratropical teleconnections, physical mechanisms predictability, and relationship with climate change (ENSO Diversity WG).

Design and promotion of PACE Fellowships
In the restructuring of US CLIVAR in 2005 the program perceived a need for the climate science community to facilitate an applications interface. US CLIVAR developed the Postdocs Applying Climate Expertise (PACE) fellowship program, inaugurated in 2007, to grow the community of experts that could bring knowledge about climate science into a decision-making context. The program, implemented by NOAA, brings recent climate PhDs together with agencies and institutions that make decisions for which climate is a factor. Through mid-2013, PACE has placed 11 climate PhDs in decision-making organizations, tackling such concerns as current and future drought, extreme rainfall events, heat waves, land use planning, risks to mountain ecosystems, cultural impacts of sea ice changes, climate change indicators in coastal marine ecosystems, climate change impacts on western snowpack, and food security in developing countries. Many alumni have taken positions subsequent to their fellowships that allow them to continue to bring their climate expertise to the interface of climate and society.

2.2 Summary

Through 15 years of coordinated research activities, US CLIVAR has played a leading role in advancing the understanding of and skill in predicting climate variability and change. The program focus has evolved from an initial em-

phasis on climate of the Americas and the role of the adjacent Pacific and Atlantic basins to a fully global research program (expanding to the Indian, Southern and Arctic basin) developing collaborations with agencies around the world. US CLIVAR research has helped establish the requisite global observing, data, and modeling systems essential for producing new and evolving climate information products to meet the growing needs of resource planners and decision-makers.

Building on the successes of the program to date, the following chapters outline the key science questions that remain, and the goals, challenges, and strategies that are needed to further advance our understanding of the role of the ocean in climate and its interaction with other elements of the Earth system.

Chapter 3
Fundamental Science Questions

One of the expected outcomes of US CLIVAR research is the development of improved predictive capability of climate and climate change for the benefit of society. In this regard, the oceans play a key role due to the large thermal inertia of seawater; as a result, the ocean provides a major long-term "memory" for the climate system, generating or enhancing variability on a range of climatic timescales. Understanding the ocean's role in climate variability is therefore crucial for quantifying and harnessing the predictability inherent in the Earth system.

Determining the ocean's role is a multifaceted problem. It begins with acquiring knowledge of the processes that govern climate variability within the oceans themselves, and extends to identifying the pathways by which the ocean modulates climate variability in the other components of the Earth system (atmosphere, land, cryosphere, etc.). The practical importance of climate predictions and projections in decision-making increases dramatically in going from global to regional spatial scales. Determining the processes by which global-scale climate variability impacts smaller-scale regions is therefore critical for developing appropriate and well-informed adaptive strategies. Finally, because of the impact of nonlinear stochastic processes, the extent to which climate variability is predictable is inherently limited, and given societal expectations, it is important to determine those limits.

To address the above issues, US CLIVAR activities focus on addressing the following science questions:

- What processes are critical for determining climate variability and change related to the ocean?

- What are the connections and feedbacks between oceanic climate variability and other components of the Earth's climate system?
- How predictable is the climate on different time and space scales?
- What determines regional expressions of climate variability and change?

The background and scope of each of the science questions, providing a rationale for the activities undertaken during the next phase of US CLIVAR, follows.

3.1 What processes are critical for determining climate variability and change related to the ocean?

Understanding what processes are critical to climate variability and change in the ocean is of importance for several reasons. Primary among these is that the first-order physical processes need to be correctly represented in climate models for improved simulation of ocean climate variability. Various physical processes determine the climate variability of oceans over different space and timescales. Examples of such processes include ocean mixing; wind driven ocean circulation; heat and freshwater fluxes at the interface of ocean with atmosphere and sea ice that control buoyancy fluctuations and buoyancy driven ocean circulation; penetration of shortwave radiative fluxes and interactions with biological processes in the upper oceans; and influence of continental shelves on boundary currents.

The fundamental physical processes that influence the ocean, in turn, determine the modes of ocean variability on various space and time scales. Oceanic eddies and waves dominate this variability on timescales of days to months and can be thought of as the oceanic counterpart of weather in the atmosphere. There are various physical mechanisms that lead to generation of ocean eddies – sudden changes in the direction of surface winds and their speed; dynamical instabilities associated with the ocean thermal fronts; and interactions between oceanic flows and bottom topography. However, such mechanisms are poorly understood.

Due to their ability to exchange energy with the large-scale oceanic state, mesoscale eddies can influence variability in the oceanic circulation and stratification on the variety of timescales. In the Southern Ocean, eddies may also transport heat, salt and biogeochemical tracers (such as carbon) poleward, and therefore, play a key role in the oceanic uptake of heat and carbon. Most climate models lack the ability to resolve ocean eddies and must rely on empirically derived parameterization schemes (Mclean et al. 2008). Despite advances in the development of such schemes, the inability of models to resolve the mesoscale eddies still counts as a major source of uncertainty in climate simulations. On even smaller scales, sub-mesoscale currents are potentially equally important for the ocean variability, but are even less well understood. Finally, diapycnal mixing associated with breaking of internal gravity waves in the interior and near rough topography likely plays a critical role in the transformation of water properties and also in the dynamics of long space and timescale processes such as Atlantic Meridional Overturning Circulation (AMOC).

On seasonal to interannual timescales, modes of ocean variability, such as El Niño - Southern Oscillation (ENSO) and the Atlantic Meridional Mode (AMM), dominate. Ocean variability on this timescale is governed by dynamical processes in the ocean and coupled air-sea interaction, including changes in surface winds. Although important advances in understanding the physical mechanisms of these modes have been made, our understanding is not complete and further research is needed (Guilyardi et al. 2012).

On longer timescales, variability in the North Pacific and Atlantic has considerable fluctuations in decadal frequencies, and is often referred to as the Pacific Decadal Variability (PDV) (Liu 2012) and Atlantic Multidecadal Variability (AMV) (Ting et al. 2011), respectively. Various mechanisms leading to decadal variability in the oceans have been posited (Soloman et al. 2011) but remain poorly understood. Basic science questions

on the role of stochastic atmospheric forcing; role of tropical-extratropical interactions and coupling with the atmosphere; and role of interactions between large-scale circulation, mesoscale eddies, and slowly propagating Rossby waves in governing ocean variability on decadal timescales remain areas of active research.

On centennial timescales, ocean variability is dominated by the global thermohaline circulation that extends throughout the water column, with one example being the AMOC (Srokosz et al. 2012). On these timescales the buoyancy driven ocean circulation is of fundamental importance. Although the difference in ocean density determined by surface buoyancy exchanges in the high latitudes is the primary factor leading to the thermohaline circulation, oceanic processes such as mixing on much shorter time and space scales are also believed to play a fundamental role.

Interactions across modes of variability can also occur over different time and spatial scales that modulate the full range of ocean climate variability. In the Indian Ocean variability associated with ENSO and the Indian Ocean Dipole influences the interannual variability of the MJO (Zhang et al. 2013). ENSO variability in the tropical Pacific also has low-frequency modulation, with some epochs having larger variability compared to other epochs (Wittenberg 2009), and models disagree on the dominant mechanisms. Higher frequency variations associated with tropical instability waves in the equatorial eastern Pacific have been hypothesized to affect ENSO variability and prediction on the seasonal timescale and have been proposed as a potential mechanism for asymmetry in the amplitude of warm and cold ENSO events (Imada and Kimoto 2012). Interactions between modes of oceanic variability across different time and spatial scales, along with the mechanisms that govern such interactions, are currently not well understood.

3.2 What are the connections and feedbacks of oceanic climate variability to other components of the Earth's climate system?

The earth system is comprised of different components - ocean, atmosphere, land, cryosphere, and ecosystems, including humans. Because of its larger thermal inertia and hence longer persistence timescales, the ocean is one of the primary sources for developing predictive capability. In realizing the predictive potential of the oceans, however, understanding the influence of oceanic variability on different components of the Earth system

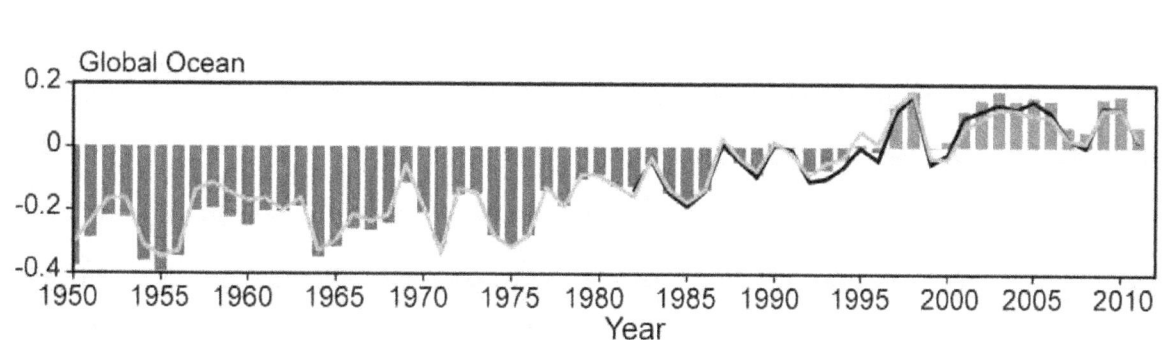

Fig. 3.1

Time evolution of annual mean of global SST anomaly from ERSST (bar) and HadISST (blue line) for 1950-2011 and OISST (black line) for 1982-2011. The slow upward trend in global mean ocean temperatures has been attributed to corresponding increases in Carbon Dioxide.

Source: Blunden and Arndt 2012, and Yan Xue.

is of fundamental importance. Understanding of the various pathways through which this influence takes place provides a basis for prioritizing targeted improvements in climate models.

Changes in variability in different components of the Earth System, attributed to the ocean, can feedback on ocean variability itself and lead to coupled interactions. Understanding different facets of coupled interactions, whereby changes in the ocean affect other components of the Earth System and in turn are affected by them, is fundamental to understanding the evolution and predictability of the modes of ocean variability. The primary mode of oceanic influence on other components of the Earth system is via changes in air-sea interaction. Variability in SST can lead to changes in the heat exchange with the atmosphere, and alter characteristics of the atmospheric boundary layer, stability, and low-level cloudiness. Such influences on the structure of the atmospheric boundary layer, its radiative properties, and the supply of water vapor can subsequently lead to changes in precipitation.

The diabatic heating associated with precipitation can be communicated to remote regions by atmospheric dynamical processes, thereby teleconnecting the influence of localized SST anomalies to atmospheric climate variability over different parts of the globe (Trenberth et al. 1998). Changes to atmospheric

circulation originating due to the oceans may eventually lead to changes in terrestrial climate and thus connect remote oceanic variability to climate variability in our back yard. Although the basic paradigm for how oceanic surface variability is communicated to remote areas is understood, the complete response of the atmosphere to geographical patterns of SST perturbations remains unclear. For example, it is unclear to what extent the unique characteristics of specific ENSO events (i.e., the magnitude and longitude of strongest SST anomalies) have a robust impact atmospheric and terrestrial variability (US CLIVAR ENSO Diversity Working Group 2013).

On the timescales of secular climate change, radiative flux imbalances due to changes in atmospheric composition are mostly absorbed by the oceans (Trenberth et al. 2009). This leads to a slow increase in SST (Fig. 3.1) that is communicated to the atmosphere and terrestrial climate. Therefore, part of the influence of the change in atmospheric composition on terrestrial climate is communicated indirectly via changes in the properties of the oceans (Hoerling et al. 2008).

Oceans are the dominant source of moisture supply and a key player in the governing the characteristics of global hydrological cycle. The rise in global mean temperature projected in conjunction with changes in atmospheric composition will result in an increase in the water holding capacity of the atmosphere. Understanding how the greater atmospheric water vapor availability may affect the intensity and characteristics of the hydrological cycle and induce variations in climatological precipitation is an essential part of understanding the response to different climate change scenarios (Liu et al. 2013). Changing ocean conditions also heavily influences where this excess moisture is distributed geographically. At low latitudes differential changes in oceanic warming could potentially produce large-scale shifts in the tropical and extra-tropical circulations. At high latitudes, where the opening of the Arctic ocean raises the potential for much larger undulations in the storm tracks, a potential for prolonged blocking events may provide new moisture sources for high-latitude continents and changing salinity patterns for high-latitude oceans.

The ability of the oceans to absorb heat and anthropogenic carbon dioxide makes them a key player in shaping the response of the climate system to the current build-up of greenhouse gases in the atmosphere. Understanding the deep pathways and their significance for climate change represents a major challenge. For example, due to its vast size and unique properties, the Southern Ocean is estimated to account for approximately half of the CO_2 and more than half of the total anomalous heat taken up by the oceans. At the same time, the Southern Ocean remains one of the most poorly sampled and inadequately represented regions in climate models used for climate projections. The challenge for the oceanographic community and US CLIVAR lies in both extending the observing systems in this and other key regions of the World Ocean and in improving climate model simulations and better understanding the physical and chemical feedbacks that may be at work.

The ocean can also interact with other components of the Earth system through the sea ice, ice sheets, and ice shelves. Climate projections indicate that later this century the Arctic Ocean will be largely ice-free during summer. How will the "opening" of Arctic Ocean change the ocean circulation and surface fluxes, and how will this influence climate variability in high latitudes? Acceleration of sea ice melt and the associated fresh water flux into the ocean may also lead to changes in the ocean density that can subsequently have a profound influence on the thermohaline circulation, but these interactions are poorly understood. Similarly, the influence of ice sheets on ocean variability show a wide range of conflicting outcomes in model simulations.

3.3 How predictable is the climate on different time and space scales?

Understanding sources of predictability on different time and spatial scales has been a difficult and challenging endeavor. However, understanding the predictability of the climate is important in order to focus efforts on the representation of those phenomena that suggest the greatest potential to add predictability. Understanding the limits of predictability will help to manage expectations of the user community regarding what predictive information science can deliver. For US CLIVAR, a fundamental issue is estimating the predictability of ocean variability on different space and timescales. On seasonal and decadal timescales, knowing the initial state of the ocean aids in oceanic predictability. For climate change projection the main driver of future conditions is exerted externally to the natural climate system, such as through the changes in atmospheric composition (Hawkins and Sutton 2009). Estimates of predictability from the initial conditions attempt to quantify far into the future the information from the initial ocean conditions that imparts information about the likely evolution of the ocean conditions (Branstator and Teng 2010). In the context of variations in atmospheric composition, the issue of estimating oceanic predictability comes down to understanding the response to the changes in radiative heating. The specific year-to-year evolution of the ocean and atmosphere circulation is considered "climate noise" relative to the changes in the climatological mean and variability characteristics of the ocean.

At present it is unclear what the true predictability of ocean variability is, i.e., what is the sensitivity to initial conditions? What are the interactions between different components of the Earth system that govern the timescale of predictability? Is the predictability of ocean variability regime and timescale dependent? What information is critical to initialize predictions for different timescales and can assist with the design of an optimal ocean observing system for monitoring and prediction purposes? Likewise, climate change projection estimates in the mean response in the oceans to variations in atmospheric composition differ widely in the current generation of climate models and need to be better constrained. Given the known limitations in climate models, it is also difficult to optimally design observational networks that can be used to test some of these questions.

Understanding the physical processes governing climate variability in the ocean and the pathways whereby the ocean

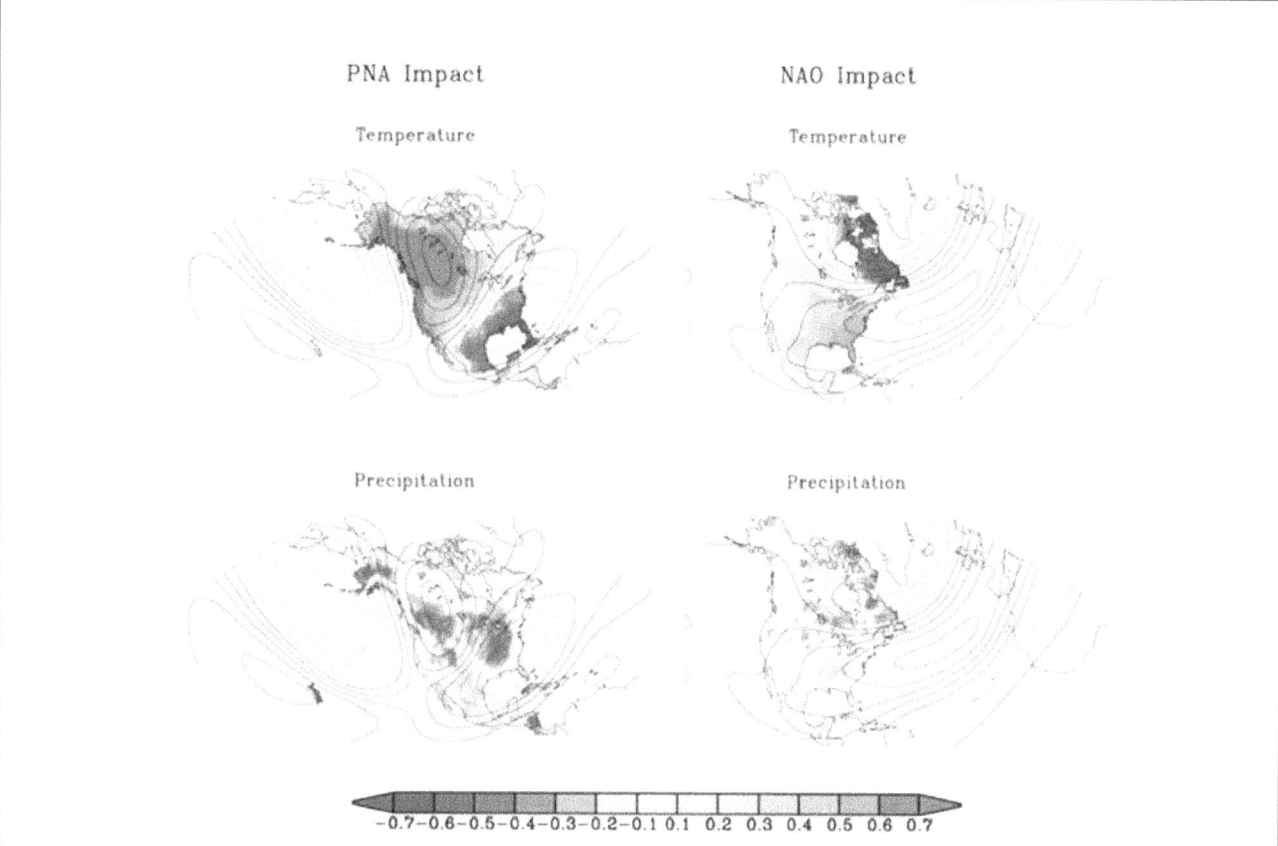

Fig. 3.2

(Left panel) *Correlation between PNA index and 500-mb height field* (contour). *The shading indicates correlations between PNA index and the surface temperature and the precipitation. The correlations are based on seasonal mean data for the period 1951-2006.* (Right panel) *Same as for the left panel but for correlation with NAO index.*

Source: CCSP SAP1.3 – Reanalysis of historical data for key atmospheric features – Implications for attribution of causes of observed change.

influences variability in other components of the Earth system are necessary but not sufficient conditions for advancing predictive capability. Another key science issue is to quantify what fraction of the variability in other components of the Earth systems is constrained by oceanic variability. Our understanding of seasonal climate variability suggests that only a certain fraction of seasonal mean atmospheric anomalies are predictable in practice. Further, not all of that realized predictability can be attributed to oceanic anomalies (Kumar et al. 2007). This predictability, or lack thereof, is not uniform but instead has a distinct geographical preference for "hot spot" regions, as well as a temporal preference for particular seasons. A similar understanding for the influence of the ocean on longer timescales, decadal for example, also needs to be advanced.

3.4 What determines regional expressions of climate variability and change?

The regional expressions of large-scale climate variability and change have societal importance in decision-making and for developing adaptive strategies to increase societal resilience. Regional manifestations of climate change are being documented based on detection and attribution studies. Examples include inhomogeneity in changes in surface air temperature including polar amplification and land-ocean contrasts, sea level, precipitation patterns, frequency and severity of extreme events, ocean acidification, and sea ice decline. All these changes are examples of spatial inhomogeneities associated with the influence of global scale climate change. Similar regional features are associated with surface manifestation of modes of climate variability, e.g., PNA, NAO (Fig. 3.2). The regional influence of changes in the characteristics of modes of variability can also be further modulated by feedbacks with local components such as those involving soil moisture, vegetation, snow cover, etc. Such interactions could be particularly important in amplifying regional expressions of modes of large-scale variability. Currently the physical understanding of the reasons behind many regional manifestations of climate variability and climate change remains an unresolved issue.

Developing a physical understanding of regional expressions of climate variability and change is important for several reasons: A sound physical understanding of the relevant climate processes and demonstration of their fidelity in climate models is critical for developing confidence in climate projections under various scenarios. Understanding the physical basis also underpins the scientific credibility of climate change. Regional manifestations of secular climate change and natural climate variability often resemble each other (Soloman et al. 2011). Differentiating between the two, therefore, is important for developing sound adaptation and mitigation strategies and is needed to gain sufficient physical understanding for decision-makers to make both short- and long-term policies in response to each. Adaptation and mitigation decisions in response to climate variability and change are made at a local scale, and a better physical understanding of regional features associated with climate change is essential to garner public support for financial resources that will be required for putting appropriate strategies into action. A physical understanding of regional expressions of climate change also plays an important role in developing appropriate regional and local indicators of climate change.

Several avenues will be pursued to investigate what processes are responsible for regional expressions of climate variability. One possible mechanism for the regional manifestation for climate change is the interaction between the relatively global signal of climate change with regional variations in the mean climate and its seasonal cycle. The mean states of different components of the Earth system also display considerable spatial inhomogeneity – e.g., the existence of cold tongue in tropical eastern Pacific; boundary currents in the oceans; meridional gradients in SST and atmospheric temperatures; distribution of mean precipitation; preferred geographical location for storm tracks. These features of mean climate are an outcome of the land-ocean distribution and annual mean distribution of solar energy. Spatial inhomogeneities in the mean climate provide a pathway for the regional manifestation for climate change. A distinct example of this is the hypothesis whereby the influence of climate change on mean precipitation is to increase precipitation in regions that have abundance of precipitation and oppositely in relatively arid regions, thus further enhancing the contrast between the wet and dry areas. Similar interactions in the mean climate are likely to play an important role in the regional manifestation of climate change and variability. A physical understanding of such interaction is of fundamental importance to society, providing an improved scientific basis for developing sound adaptation strategies.

Interaction between weather extremes and climate change offers another pathway for regional manifestations of climate change. In the present climate, weather extremes such as hurricanes and extratropical storms, tornadoes, heat waves, and droughts have geographical preferences and are controlled by features of mean climate and by modes of climate variability. How weather extremes will be affected by future alterations in climate variability and change is an important area for improved understanding.

Chapter 4
Goals

US CLIVAR has articulated the following set of goals to address our mission:

- Understand the role of the oceans in observed climate variability on different timescales.
- Understand the processes that contribute to climate variability and change in the past, present, and future.
- Better quantify uncertainty in the observations, simulations, predictions, and projections of climate variability and change.
- Improve the development and evaluation of climate simulations and predictions.
- Collaborate with research and operational communities that develop and use climate information.

The goals draw attention to the critical issues of scale, modeling, communication, and the characterization and interpretation of uncertainty. Goals 1-4 are similar to ones pursued in the previous program (Chapter 2). Goal 5 is new in that it stresses interactions of US CLIVAR with other research communities of the Earth System and with the applications community. As such, there is a close connection between the goals and the Fundamental Science Questions (Chapter 3). Although the goals do not address any of the questions directly, they focus on broad areas of understanding and capability that are applicable to the complete set of questions.

4.1 Goal 1: Understand the role of the oceans in observed climate variability on different timescales.

The oceans are known to impact the climate and its variability on a wide range of timescales. From short timescale effects of the ocean on weather, such as the role of the Gulf Stream influencing the path and intensity of Hurricane Sandy, to the thousand-year timescale of the meridional overturning circulation, the ocean acts on a wide range of timescales as a repository of heat, energy, and gases important to the climate system as a whole. The warming of the ocean surface in recent decades has also been an important reservoir for global energy, a source of steric sea level rise, and a key factor in the reduction of summertime sea ice in the Arctic. US CLIVAR coordinates research spanning all of these timescales.

Climate variability occurs in all components of the Earth system, the oceans, atmosphere, and cryosphere. As outlined in the Fundamental Science Questions of Chapter 3, US CLIVAR seeks to understand where this variability originates in the ocean, and how the ocean variability interacts with the other components. While processes and feedbacks may cross temporal boundaries, separating by timescale provides a convenient framework for elucidating science issues in a succinct manner.

Intraseasonal variability

The MJO has been linked to the variability at intraseasonal timescales over the globe including the tropical cyclone genesis (e.g., Maloney and Hartmann 2000), mid-latitude weather and climate (e.g., Higgins et al. 2000), and high-latitude circulation (e.g., Zhou and Miller 2005). The MJO also affects the variability and prediction of the climate at longer timescales such as the El Niño Southern Oscillation (ENSO) (e.g., Kessler and Kleeman 2000).

Simulation of the MJO has improved considerably in climate models leading to improvements in its prediction skill, and these are prior efforts whereby US CLIVAR has made substantial contributions. Observations of exchanges between the Indian Ocean and atmosphere, cloud and convective processes, and fundamental dynamical and modeling improvements have all contributed to these advancements (e.g., Zhou and Murtugudde, 2009; Zhang et al. 2013). However, despite successes in simulating the MJO, substantial problems still remain, particularly with the understanding and simulation of tropical convection, its interaction with the MJO variability (e.g., Subramanian et al. 2013), and the initiation and propagation of the MJO and its interactions with Maritime Continent (Fu et al. 2013, Zhang et al. 2013).

Along with the MJO, other modes of variability associated with the oceans are important on intraseasonal timescales. These include: blocking in extratropical latitudes which is sometimes connected to western boundary current variability; precipitation, soil moisture, and snow anomalies in determining terrestrial climate; etc. (see Lau and Waliser 2012 for a review).

Seasonal to interannual variability

The difference between maritime and continental climates demonstrates the role of the ocean in climate variations. Coastal and island regions have dramatically smaller seasonal variations than dry regions separated from the ocean, such as high mountains and subtropical deserts. The moderation of temperature results from the exchange of heat and moisture between the atmosphere and ocean. Sea water stores much more energy for a given change in temperature than air, and the evaporation of sea water requires a great deal of energy. Furthermore, even over land, atmospheric water vapor strongly affects the balance of energy from the sun and energy lost to space. Thus, the presence of seawater and water vapor in a region strongly affects seasonal to interannual climate.

Advances in seasonal prediction and operational forecasting capabilities are examples of societal benefit gained from a collaborative research effort spanning several decades. The dominant source of predictability on seasonal to interannual timescale is associated with slow variations in SSTs, particularly those associated with ENSO. Seasonal predictability that arises from slow variations in SSTs—soil moisture; snow; sea ice extent and concentration; tropospheric-stratospheric connection—have also been explored as possible sources of predictability. The fundamental tenet of mechanisms controlling seasonal predictability, therefore, arises from the influence of slow variations in boundary exchanges influencing atmospheric and terrestrial variability.

Monsoons are a key source of moisture sustaining agriculture, water supply, and therefore habitability. Better prediction and resource management requires sustained observations, as well as better process understanding and computational capabilities. The TAO/TRITON (Pacific), RAMA (Indian), and PIRATA (Atlantic) arrays (section 2.2 above), as well as other remote and *in situ* oceanographic data provide both context and initialization data for monsoon prediction. Interannual variability (e.g., ENSO, Indian Ocean Dipole) of each of the tropical basins influences monsoons and droughts throughout the region (e.g., Ummenhofer 2009).

Analyzing the slowly varying exchanges between Earth system components provides a framework to identify important processes that govern seasonal variability. Some processes within a component of the system affect the variability, and sometimes it is the exchanges across the interface between two components that lead to variability. For example, oceanic mixing and inter-ocean or tropical-extratropical exchanges are important in governing ENSO variability, and improved monitoring and modeling of these processes has led to the improvement in understanding and prediction of variability. The air-sea exchange of heat, vapor, and momentum across the ocean and atmosphere interface also plays an important role in evolution of the seasons, ENSO, and other variability on these timescales. Mechanisms such as the influence of remote ocean basins via atmospheric bridges (Alexander et al. 2002), seasonal footprinting (Vimont et al. 2003), and the influence of seasonal to interannual SST variability on precipitation, clouds, and radiative fluxes at the ocean surface are all important for overall climate sensitivity.

The oceanic state strongly affects seasonal to interannual variability, such as ENSO. Changes to SST, thermocline depth, air-sea feedback strength, and the meridional reach of oceanic variability through large-scale planetary waves and the subtropical cell have all been shown to affect ENSO variability in models. However, not all models agree as to how strong these changes will be (Collins et al. 2010). Climate variability in regions outside the tropic may similarly be sensitive to the oceanic state. Therefore, understanding and modeling of the processes that can affect these properties of the oceanic state, such as upper ocean mixing, clouds, and air-sea exchange of heat and momentum, are important for understanding and modeling interannual variability.

Oceanic variability can also interact with and modulate other slowly varying components of the Earth system. For example, ENSO and the NAO affect sea state properties in regions where these effects can be felt globally through atmospheric and oceanic pathways. These short- and long-range effects can enhance or inhibit other modes of variability on longer and shorter timescales.

Decadal variability

The value of improved near-term (0 to 2 year) regional climate information for society has prompted considerable research in the field of decadal climate predictions (approaching 10 years and beyond). The basis for decadal prediction is the existence of modes of variability with long timescales – e.g., Pacific Decadal Variability (PDV); the Atlantic Multidecadal Variability (AMV) and its counterpart in ocean currents the Atlantic Meridional Overturning Circulation (AMOC). In each of these variability examples, the ocean plays a dominant role. Additionally, ocean processes (e.g., overturning and gyre variability, subtropical cells, Rossby waves and mesoscale eddies, and advection of temperature/salinity anomalies) may provide a basis for oceanic anomalies that persist for decades.

A variety of different mechanisms have been suggested to explain decadal climate variability (Frankignoul 1985). One hypothesis argues that midlatitude SST variability is caused by random atmospheric surface forcing being smoothed out by the ocean to produce lower frequency variability. Alternative explanations include tropical forcing associated with ENSO, extratropical atmospheric stochastic forcing, changes in the North Pacific oceanic gyre and the Kuroshio Extension, and the reemergence of SST anomalies due to the strong seasonal cycle of oceanic mixed layer depth. New discoveries continue (e.g., Newman et al. 2003, Eden & Willebrand 2001, Böning et al. 2006). US CLIVAR will foster better understanding of these mechanisms and assessment of their utility for predicting decadal variability.

The current generation of climate models does not consistently replicate the frequency spectrum associated with different modes of decadal variability (Furtado et al. 2011). PDV, for example, is not well simulated (Deser et al. 2012a). Care is required in interpreting the observational record, which is limited in assessing decadal variability because we have not monitored many of these variations (Wittenberg, 2009; Stevenson et al. 2010, 2012). Much of the dynamical response of the South-

ern Ocean to decadal wind changes relies on the response of unresolved mesoscale eddies, so the present coarser resolution models, which do not resolve the eddies, may have incorrect sensitivity. It is suspected that inaccurate variability stems from the difficulty in accurately modeling many processes in models. Particularly troublesome are the representation of mixing in the upper ocean, clouds and aerosols, future ozone concentrations, modeling regions of weak vertical stratification in the ocean, and ocean-sea ice interactions (e.g., Booth et al. 2012).

Centennial variability

Earth has warmed over the last century owing to increasing concentrations of greenhouse gases. Further warming is likely to continue over the next decades and centuries even with attempts to mitigate anthropogenic impacts. To develop mitigation and adaptation strategies and to inform the decision-making process, sufficiently accurate, credible, and understandable climate information at global and regional scales will be required. However, uncertainty remains about the magnitude and rapidity of the warming in response to a given increase in greenhouse gases, and what types of changes in climate variability will accompany such warming.

Climate sensitivity quantifies the warming resulting from a given radiative forcing. The most common definition of climate sensitivity is the equilibrium mean surface warming in response to a doubling of atmospheric carbon dioxide concentration, but other measures are also used. The accuracy and precision of these estimates have not improved markedly over the past decades. Although model simulations have improved, certain biases continue to confound attempts to understand how changes in atmospheric composition relate to circulation and regional climate change. It is of great concern that model adjustments to "tune" poorly-known parameters continue to negatively influence our assessment of climate sensitivity (e.g., Kiehl 2007). The uncertainty in climate sensitivity also limits prediction of many aspects of climate change, such as the occurrence of heat waves, droughts, and floods. Quantifying the uncertainty is crucial in developing reliable cost-benefit analyses of mitigation and adaptation strategies.

Because the ocean is slow to warm with added thermal energy, much of the past energy imbalance caused by increased greenhouse gases has gone toward raising the ocean temperature moderately. Monitoring these changes and understanding how they are distributed through the ocean is a key part of the

Earth's climate sensitivity. US CLIVAR's emphasis on ocean observation and processes involving the ocean and its exchanges with the atmosphere, land, and cryosphere mean that US CLIVAR can play a unique role in US efforts in this monitoring and understanding. Changes of state of water—melting of land and sea ice, freezing of seawater, and evaporation and precipitation—also contribute significantly to the Earth's energy storage and transport, as well as to processes of the cryosphere, atmosphere, and ocean.

Centennial climate variability shares processes and timescales with global change, but the observational record is too short to understand centennial variability in the absence of global change. It is unclear to what extent different centuries vary, and observations on these timescales are limited to historical reconstructions or indirect evidence (paleoclimate records of sediments, ice, or biological traces that are related to or "proxies" for the desired climate variables). US CLIVAR seeks to improve the accuracy of models and theory so that these tools can be used to understand variability and change on the longest timescales. Integrated assessment with observations, modeling, and collaboration with paleoclimatology efforts can help to push the boundary of what is known about climate variability and change toward the centennial timescale.

Interactions across timescales

Exchanges between Earth system components are also influenced by variability on different timescales. A familiar example is the influence of MJO variability in determining the initiation, termination, and the amplitude of ENSO variability. An oceanic example is the effect of variability of North Atlantic Deep Water formation on the Atlantic Multidecadal Oscillation and AMOC. Climate variability on longer timescales—interannual, decadal, centennial, or anthropogenic warming—can modulate the characteristics of weather and seasonal variability, and some of the changes in the past have been severe enough to have altered human civilizations. Because of the direct societal impact through agriculture, water availability, and catastrophic events, understanding and predicting the consequences of these connections would be beneficial. Progress has been made in understanding physical processes that govern the evolution of slowly varying boundary exchanges and simulating and predicting their evolution in climate models, but advancing our understanding of these important physical processes and interactions among them is still a key goal of US CLIVAR.

Interactions across spatial scales

There is often a connection between time and spatial scales of variability. Typically, the variability of the globe or large regions is slow, and smaller regional variability is faster. However, this is not always the case, and certain small regions may have important low-frequency variability. For example, Atlantic Multidecadal Variability most strongly affects a relatively small region around the Labrador, Greenland, Iceland, and Norwegian Seas. Droughts and monsoons are similarly regional in scope. Seasonal variability, on the other hand, occurs on a relatively short timescale but is global in extent. While the scale of problems considered by US CLIVAR naturally extends to the global scale, regional variations and "climate downscaling" to regions is a part of understanding variability on various time and spatial scales.

4.2 Goal 2: Understand the processes that contribute to climate variability and change in the past, present, and future.

An important step in understanding climate variability is the identification of contributing processes. US CLIVAR coordinates sustained observations, process studies, modeling, and theoretical efforts to help identify and quantify climate variability and the processes that cause it. Only through a synthesis of understanding of all of these important processes can we hope to quantify the possible range of future outcomes and their sensitivity to human or natural perturbations that may occur. Examples of variability featured in US CLIVAR-facilitated research spanning multiple timescales are given in section 4.1 above. Here some of the processes and phenomena understood to contribute to this variability of the climate system are described.

Physical climate processes

Over the whole planet, heating by the sun is very nearly balanced year-by-year by the energy lost to space, which is affected by the amount of greenhouse gases (such as water vapor and carbon dioxide) in the atmosphere. Latitude by latitude there is an imbalance between incoming solar and outgoing energy to space. In the simplest terms, the role of the fluid components of the Earth system—atmosphere and ocean—is moving excess solar energy at in the Tropics toward the poles where it is emitted back to space (Trenberth et al. 2009).

The atmosphere and ocean rebalance the energy directly through fluid motions (convection) of the atmosphere and oceans and indirectly by latent heat release. Weather and climate variability are consequences and agents of this redistribution. The oceans play many roles in the Earth's energy balance: (1) as a source of the primary greenhouse gas (water vapor) and water for cloud formation; (2) as a source and return path for latent energy (water vapor); and (3) as a poleward transporter of energy through warm poleward and cold equatorward flow. Roughly, the atmosphere, ocean, and water cycle (vapor & liquid) play equal roles in the redistribution of energy from the equator to the poles.

Key processes transport, stir, and mix warm and cold or dry and moist fluid elements. Convection also affects clouds and water vapor, thereby affecting the energy balance of the Earth. Vertical convection of warm, moist air and precipitation importantly connect the higher altitudes with the surface, just as convection in the ocean brings cold, salty water to depths where it fills the ocean abyssal basins. Many atmospheric and oceanic properties as well as climate variability rely on communication with the boundary between these fluids, so the mixing and exchange across the boundary layer and vertical convection is a prerequisite for accurate simulation and prediction. Atmospheric convection affects water vapor, aerosols, and clouds, which in turn affect the radiative components of the energy balance. Atmospheric convection and precipitation is also the dominant source of freshwater for the oceans. Better representations of mixing and convection—oceanic and atmospheric—will deliver more robust climate models and projections of climate and climate variability. The details of the atmospheric and oceanic boundary layers and convection, the sources and fate of aerosols, and the cloud microphysical processes that control the clouds, aerosols, and water vapor concentration are not well-represented in climate models. US CLIVAR facilitates many process studies and model development efforts to rectify this problem (Section 6.2).

The ocean has a profound impact on the water cycle. Evaporation from oceans is the dominant source of atmospheric water vapor. The oceans are an active reservoir of water-their temperature and circulation strongly affect the degree to which evaporation and precipitation occur. Likewise, these processes affect the ocean stratification and circulation through salinity and latent heating. Global warming also results in a moister atmosphere that traps more of the outgoing energy and warms the planet further. This water vapor feedback amplifies global climate variability and change by roughly a factor of two. Precipitation over land and sea is also affected by the global moistening, influencing droughts, floods, and extreme storms.

Many important sources of climate variability stem from regional or episodic variations in convection or mixing. Examples include the MJO, which organizes an atmospheric convection anomaly, and the formation of deep water in the Labrador and Irminger Seas, where variability in small-scale convection affects much larger phenomena such as ENSO and the AMOC. Identifying the root cause of the variability leads to the study of how, why, and when these variations occur. Convection occurs on small scales, just 1-10 kilometers in the atmosphere and even smaller scales in the ocean. Mixing processes are smaller still, sometimes turbulent eddies just meters across contribute importantly. Small openings in sea ice, called leads or polynyas, greatly affect the rate of air-sea and ice-sea exchange and convection. Ocean warming of ice shelves may play a key role in the variability and breakdown of these features. Variations in the larger atmosphere, cryosphere, or ocean can provide conditions that accelerate or decrease the rate of these small-scale processes. Identifying and representing these processes and their variability is a key challenge for US CLIVAR. Through improved process-level understanding, better reconstructions of past events, better present monitoring, and better projections for future outcomes will be possible.

Biogeochemical processes

Not all climate variability is encapsulated by physical climate variables. Equally important is the variability of the biogeochemical cycles of the Earth. The ocean plays a major role as a reservoir in many of these cycles, such as the carbon, sulfur, and nitrogen cycles. Both biological and inorganic (gas solubility and seawater chemistry) processes are important in determining the cycles of these natural chemicals. The partitioning of anthropogenic CO_2 between the atmosphere and the ocean has profound implications for future warming and for ocean acidification. The addition of nutrients like fixed nitrogen and phosphate to the ocean causes increased biological productivity. Sometimes the productivity is excessive, causing anoxia or harmful algal blooms, with societal impact for coastal communities and fisheries. The circulation of the oceans and its variability both strongly affect biogeochemical cycles, as the solubility, storage, and transport of these chemicals is a function of the physical state of the seawater and its motion. Changes in the biogeochemistry of the ocean in turn affect the physical state by altering turbidity, aerosols, clouds, and surfactants that affect air-sea exchanges.

Crucial processes contributing to ocean biogeochemistry affect the exchange of chemicals and water between the air, sea, ice, land, and the Earth's crust. The surface boundary layers of the atmosphere and ocean are controlling factors in how these exchanges occur. Ocean biology, especially photosynthesis, also occurs primarily in these boundary layers where light is plentiful. Then, processes such as convection and oceanic upwelling and atmospheric subsidence connect the deep ocean and troposphere to the boundary layers, providing nutrients and moving dissolved compounds into deeper waters. Finally, on the longest timescales, storage in abyssal waters of the ocean, biological and geological sedimentation and seafloor burial, formation of seafloor, volcanism, and the subduction of deep ocean seabed into the mantle at trenches decide the ultimate balance of the Earth's biogeochemistry. Paleoclimate reconstructions have helped us to understand how processes and budgets on different timescales contribute to past, present, and future variability.

Role of observations and models for understanding processes

Observations that are used to reveal, monitor, reconstruct, or project changes in these key processes are at the foundation of US CLIVAR's work. Process studies help to develop understanding and monitoring systems, which then can be used to study variability on longer timescales. From observation to understanding and modeling and then on to monitoring and prediction, US CLIVAR helps to facilitate and focus the scientific effort (Section 6.1). Long- term retrospective analyses (or reanalyses) provide observationally driven, globally continuous datasets which also include representation of the physical processes that are difficult to observe. While background model uncertainty plays a role in the reanalyses, the number of different reanalyses is growing, allowing the uncertainty to be quantified in an ensemble sense. In addition, research progresses on integrating Earth system components in reanalyses (e.g., ocean, atmospheric, land, cryosphere, chemistry, and aerosols), which should enhance the understanding of the system interactions.

As we cannot freely experiment with the Earth, models play the role of laboratory experiments in Earth science, and so a goal of US CLIVAR is to develop new modeling capabilities to help understand these processes and their role in climate change and variability. Sometimes observing a process is not sufficient to determine its role in the climate system—what would happen if it were much weaker or stronger? Numerical models are able to address such questions directly. "Virtual Earth" experiments

quantify the impact of a particular process by changing it; how does a nearly identical Earth with only that process altered behave? A second method of understanding key processes through modeling is the use of high-resolution models to capture the small or fast phenomena directly. Typically, such phenomena are not easy to observe accurately enough to diagnose their effects on climate variability of energy, water, or biogeochemistry.

As we discover new physical processes and assess their roles in climate variability, targeted observations and strategies are needed to track their evolution, variability, and response to climate change. Only a careful and thorough collaboration between the monitoring and science communities can such tools be successfully developed and implemented. Only a clear understanding of the processes—guided by process studies using models, observations, and theory—will lead to robust improvements in simulation and prediction of climate variability. These sustained observational systems will be our eyes on the world to measure and predict variability of key climate processes.

As key processes are identified and sustained observations are put in place, it is crucial for US CLIVAR to support a continued and integrated assessment of these processes. Enhanced observational arrays and improvements to models and theory will help to understand and anticipate variability in these processes and thereby variability of the climate as a whole. Continuous monitoring is also a necessity for understanding both the past and the present variability in a proper context for projecting future outcomes. Integrated assessment is particularly important for improving the representation of these processes in models. On the longest timescales, integrated assessment between paleoclimate and modeling communities can be extremely valuable. Often processes are well-understood from an observational or theoretical standpoint, but unless those scientists work closely with modelers these processes will not be adequately represented in climate models. Chapter 6 details many of the approaches that US CLIVAR uses to facilitate the transition from process study to monitoring.

4.3 Goal 3: Better quantify uncertainty in the observations, simulations, predictions, and projections of climate variability and change.

What will the future climate be? It is now common to project and simulate future climates to assess the likelihood of variability and change. All groups within US CLIVAR are involved in aspects of these projections, be it through modeling future scenarios or past variability, observing the present, or interpreting variability and change of past climate. To put our future projections of climate and climate variability in context, the uncertainty of these projections needs to be assessed.

Understanding and quantifying error sources

A gold standard is to have quantification of uncertainties. It is important to recognize that some uncertainties are intrinsic due to tiny inaccuracies in estimating the state of the Earth system that are amplified by its chaotic nature (the "butterfly effect"). Other uncertainties, called model "errors" or "biases", are decreasing as we improve simulations and better understand how to represent key processes. US CLIVAR Goal 3 seeks to quantify both categories of uncertainty, while other Goals (2 & 4) seek to reduce uncertainties of the second kind.

First consider the errors in simulations and projections that can be reduced, that is, that are due to our lack of knowledge or limited computing power. Identifying such errors requires careful and comprehensive comparison to well-calibrated and sustained observations of key processes. Through such observations, we can quantify the mean climate state, its variability, and the frequency and intensity of climate extremes. By comparison, model errors may be revealed in reproducing such statistics, and then attention can be paid in model development to making the models more realistic in terms of their climate and variability (Hawkins and Sutton, 2009).

Contributions to uncertainty from unresolved processes

As discussed in Goal 2 (Chapter 4.2), it is likely that better modeling of the climate and its variability will require a better representation of the key processes that contribute. In numerical models, some processes are "resolved," which means that with present computing power the time and spatial scales of those processes are directly simulated with only numerical approximations being a source of error. Many other processes are too small or too fast to be resolved, so they must be "parameterized." Often the error associated with each parameterization or numerical technique is estimable, but how these errors accumulate through the complex interconnections between climate model components is not easy to estimate. Thus, quantification of the combined effect of these errors on climate projections is supported by US CLIVAR. A more difficult task is identifying how much error results from processes that are unresolved but not yet parameter-

ized, which can only be carried out by comparison of the simulated climate to high-quality, sustained observations of the Earth.

While many of the key climate processes that are well understood are already parameterized or resolved in present climate models, some processes, such as internal and surface ocean gravity waves, have yet to be parameterized fully even though these processes have been studied at a process level for decades. Land ice is another example of a system that is not modeled in process-level detail in most climate models—as we become more interested in centennial variability this gap will be corrected. It is natural for such oversights to be recognized as other processes are better parameterized or resolved in improving models or as our interests extend to longer timescales. If, for technical reasons, neglected processes cannot be easily parameterized, then it is important and usually easier to quantify the error associated with neglecting them as a first step.

As computational power increases, many processes will move from being parameterized to being resolved. Present examples are the atmospheric and oceanic mesoscales, which can be resolved in high-resolution prototype calculations but not yet operationally in global models. These costly calculations can be used as a basis for quantifying the errors in standard-resolution calculations. Additionally, US CLIVAR can encourage the deployment of better computational capability and the development and implementation of more efficient numerical methods. These steps will speed the process of transitioning processes from being parameterized to being resolved.

Uncertainty in regional and downscaled predictions

High-resolution or down-scaled models may be necessary to address regional impacts of climate variability. The fundamental science question (Chapter 3.4), "What determines regional expressions of climate variability and change?" provides a challenge to modeling, as global models capable of focusing on the global variability are less well-suited to assess regional impacts of the variability. Thus US CLIVAR collaborates with many groups that participate in regional modeling and observation. However, it is important to note that large regional variability may not always be generated by large global variability—sometimes regional effects of one type of variability are excessive while other regions respond more to other types of large-scale variability. For this reason, assessing the regional impacts of climate variability is an ongoing challenge and an outstanding fundamental science question.

Development of new approaches for uncertainty quantification

Quantification of model errors is not simple, as the sources and measures of error are not always clear. Weather in climate models will always differ from the weather in the real Earth, but perhaps the climate change, variability, and extremes may be correctly modeled. US CLIVAR will encourage the development of sound measures for evaluation of models against observations and against higher resolution calculations so that errors can be properly quantified and reduced.

Quantification of uncertainty and reduction of error is crucial and may require collaboration between different groups. The Model Intercomparison Programs (AMIP, OCMIP, CMIP) are an excellent example of this process (Chapter 2.2), and the Earth System Grid Federation (ESGF) is attracting new datasets to support model intercomparisons. US CLIVAR strives to identify and enhance pathways for quantifying and understanding present model biases and errors. Where possible, these errors can be reduced with improved models (Goal 4).

Reanalyses products offer global observationally-based data for dynamics and physical processes, yet still have uncertainty related to the background model and assimilated observations. Reanalyses will begin to provide their forecast difference from the assimilated observations (observation minus forecast) alongside the assimilated observations. This output provides information for researchers as to where observations exist and which observations influence the assimilation most. Observationalists may also be able to use this output to determine errors in the assimilated observations. With increasing numbers of reanalyses becoming available, discrepancies among the reanalyses can also be quantified by comparison. This will become increasingly important in understanding the uncertainty of integrated Earth system analyses.

Many of the efforts within US CLIVAR, especially the Process Studies Panel and Climate Process Teams, focus on a few components of the Earth system at a time. Quantification of the errors and uncertainty of individual processes is insufficient to capture the error and uncertainty of a whole projection or simulation. Processes not yet identified as contributing, and the amplification of errors and uncertainties through coupling of different components within the chaotic Earth system can both contribute to additional errors and uncertainties not easily identified by process-level study. A holistic assessment of the errors and uncertainties of a climate simulation or projection is therefore also needed.

US CLIVAR aids development and adaptation of tools to use multiple models and ensembles to assess predictability, skill, sensitivity, and uncertainty in mean state, variability, and extremes. Many techniques required to effectively compare a disparate set of simulations have been developed, often for use in weather forecasting rather than climate projections. US CLIVAR can make progress with joint workshops, working groups, and data standards that will ease the transition of existing technology and development of new technology for this complex task.

The collection, identification, sharing, and development of relevant datasets, state estimates, reanalyses, and reconstructions plays a key role in the evaluation of simulations and the development of statistical measures of uncertainty. Sustained, consistent data is our primary way of understanding the climate variability that we have witnessed. Gaps in data and lack of unusual or extreme occurrences greatly inhibit our ability to predict and simulate future related events.

A challenge in public communication is relating levels of confidence and uncertainty. Challenges in collaborating with research communities that develop and use climate information include communicating predictability and skill of simulations and projections and managing uncertainty in their risk and impact analysis. A key role for researchers is development and documentation of robust metrics and products from models, which can be used to inform the public and evaluate models with data. Robust metrics are better suited to explanations or measures of uncertainty. For example, the first day of snow, how many days before that snow melts, or the seasonal accumulation of snowfall are all societally-relevant measures of snow. However, they differ widely in their degree of predictability and uncertainty.

Different timescales and regions should have different metrics of phenomena, data quality, predictability, and uncertainty that are the most relevant and robust. US CLIVAR will work to develop a dynamic envelope of metrics of data quality, predictability, and uncertainty spanning intraseasonal to centennial and regional to global scales.

Simulations are a key tool in assessing uncertainty and predictability. Despite the persistent biases and errors in global climate simulations and the variety of issues in using more simplified or empirical simulations, they remain our best tool for studying variability beyond what observations and theory can assess. Large-scale climate variability in models should be studied carefully and evaluated carefully for use in quantifying possible future states and variability. Synthetic observations can be made in these simulations to optimize locations for experiments and monitoring stations. A diversity of models is also a useful measure of their uncertainty, and when they agree it enhances our confidence in their projections. Maintaining a broad suite of approaches to modeling climate variability should be encouraged by US CLIVAR.

4.4 Goal 4: Improve the development and evaluation of climate simulations and predictions

There is a variety of climate simulations used scientifically and operationally by the scientific, industrial, and governmental communities. Some of these simulations are the result of simple models, developed by small groups, some are highly complex, with person-centuries of development time.

Forecasting, prediction, projection, and experimentation: simulation strategies

A forecast is a best estimate of what is to come in the near future. In numerical modeling, a weather forecast includes all of the variables of consequence to society: temperature, precipitation, pressure, winds, etc. A forecast is usually generated by expert analysis of predictions, which are probabilistic determinations of the possible future states of the Earth or a region and their likelihood. In making a forecast, an expert assesses and synthesizes the complexities of the different outcomes believed to be possible based on the predictions for a broad audience.

Numerical model simulations are often employed in making predictions—often many simulations are grouped together into an ensemble to help determine many possible future outcomes. The behavior of many components of the Earth system, such as the atmosphere, is chaotic, and the "butterfly effect" is one consequence. For this reason, it is presently believed that individual simulations of weather are less useful than an ensemble of simulations, allowing for different initial conditions and variability all constrained to agree appropriately with any observations. Alternatively, a "stochastic" model directly simulates the uncertainty associated with chaos in a single simulation. In either case, the chaotic behavior of the Earth system means that the short-term variability, or weather, in these models begins

to lend such large uncertainty to predictions that after about two weeks, all evidence of the initial conditions is lost. Thus, weather simulations, and the predictions and forecasts that rely on them, are limited to within this two- week window. During the two-week window, however, weather variability typically dominates variability due to climate processes, such as the small planetary energy imbalances due to trends in greenhouse gases or the remote effects of El Niño. For this reason these models tend to neglect the processes needed for studying variability at longer timescales in favor of focusing more on improving processes important for the weather forecast timescales.

Climate models are used to understand the behavior of the Earth system as a whole on much longer timescales, from seasons to millennia. The work of US CLIVAR often focuses on these longer timescales. While the weather cannot be forecasted or predicted so far in the future, the probability of possible weather conditions or other statistics averaging over weather conditions, such as global mean temperature, can be projected over longer timescales. A projection is therefore less specific than a prediction or forecast, and only some variables may be usefully projected. Over such long timescales there are many natural and societal uncertainties outside of the initial conditions, such as human choices in future emissions of aerosols and greenhouse gases or the occurrence of future volcanic eruptions. Projections try to span these uncertainties by using a suite of future "scenarios." As in weather predictions, ensembles of simulations are often used to estimate the probability of various outcomes under each scenario.

Numerical models can also be used to understand a particular process or component of the Earth system. In this setting, "experiments" are carried out in a context removed from the present or past state of the Earth, with a goal toward testing hypotheses about how the system itself works. Models used in this context are sometimes as complex as those used in forecasting and projections, but sometimes they are much simpler and designed just to focus on a particular process or behavior. Model perturbation experiments are an important part of understanding the components and connections between pieces of the Earth system.

One topic of present research is how best to initialize models for projections of weather and climate. For short-term weather forecasts, a "nowcast" assimilation of interpolated ocean and ice observations is typically used, along with an ensemble of atmospheric initial states to span the possible weather. However, for longer timescale projections, where decadal or multidecadal variability may play a role, the state of the ocean at the beginning of the projection may affect the projections for some time. While the present ocean observing system measures the present state of the ocean above 2000m depth continuously and nearly globally, trends in modes of decadal and multidecadal variability are not well known. Thus, climate projections are often partly initialized with climatology rather than direct observations. Ongoing experimentation at modeling centers, involving various forms of ocean spin-up or ensemble strategies, is likely to yield better projections and estimates of medium-term uncertainty in climate projections.

Importance and development of models

The quality and validity of the results of numerical experiments are predicated on the accuracy and appropriateness of the models used and the observations against which they are validated. In data assimilation, the background model is assumed to be unbiased. Typically, however, this is not the case and so model bias can contribute to uncertainty in analyses and reanalyses. Simple models can be written and evaluated, even by an individual in some cases, but complex models such as those simulating the entire Earth system or assimilating all available data for weather forecasts require a community-wide effort to develop and evaluate. Aiding in this process will help reach US CLIVAR's Goal 4.

Improved numerical treatment of resolved processes has been a focus of applied mathematics since even before electronic computers existed. However, new developments applicable to climate simulations continue to be discovered, and part of US CLIVAR's Goal 4 is to incorporate these numerical techniques. For processes that are unresolved, a parameterization is needed for those processes that contribute to climate and climate variability identified in Goal 2. Parameterization development and testing requires a deep understanding of processes that can be gained through a combination of process observations, process-resolving models, and theory. Dedicated collaborations among climate scientists—modelers, observationalists, and theoreticians—are therefore needed to realize or improve each parameterization or numerical improvement. Quantification of how uncertain each parameterization and the parameters that go into it are is a key part of identifying potential model error.

The models we presently use as tools differ in the degree to which they incorporate data and observations and the degree to which they conform to scientific principles, such as conservation laws and kinematic constraints. A key development need

that US CLIVAR will address is how data can be increasingly incorporated into models that lack it, and how models that incorporate data in an empirical way can be improved with scientific principles. US CLIVAR has fostered progress on these fronts by bringing together the scientists developing data-based and principle-based models to work together on particular processes or topics. Improving models to be both data-constrained and scientifically-consistent will lead to improved monitoring, understanding, and prediction of climate variability.

An important aspect of model development is model evaluation. Part of model evaluation is comparison to data, with appropriately developed diagnostics. Many branches of US CLIVAR are involved in collection, curation, and diagnosis to improve this stage of model evaluation. A new branch of statistics, uncertainty quantification, involves the quantitative assessment of uncertainties in various applications. Of course, statistical measures of uncertainty alone do not improve models. However, since models rely on many unknown parameters and parameterizations, and since the models themselves are complex systems of many different interconnected pieces, uncertainty quantification can aid in accelerating the pace of model development by better understanding what controls model behavior. Uncertainty quantification together with good diagnostics will improve forecasts and projections (e.g., Murphy et al. 2004).

Even when data is conveniently formatted and used in models, the quality, longevity, and continuity of climate records is a problem for model development. If datasets of differing quality are assimilated into a model without regard to which is the more accurate, then the model as a whole will suffer. Unified and accurate assessment of data quality is critical. The quality of model and simulation output, be it a prototype or well-evaluated release code, also requires thoughtful communication to accommodate the needs of diverse user communities. Reviewing procedures and sharing best practices for the reporting of data, simulation output, and even model manuals is a way that US CLIVAR can contribute. Finally, climate variability is by definition a set of phenomena that change over long timescales. The datasets needed to constrain models on these timescales must be commensurably long and of high quality. US CLIVAR must continue to support the continuity of long climate records of high quality, so that these data can be used to improve and evaluate models.

There are few scientists who understand where models are weakest and most likely to produce spurious or questionable answers better than developers themselves. Yet developers are often not included in the design of the observational networks and process studies that are used to evaluate model weaknesses. US CLIVAR develops pathways, such as Climate Process Teams, where modelers can play a more active role in guiding observations toward diagnosing key parameters that can be best used to constrain and evaluate model behavior (e.g., Cronin et al. 2009).

Climate and weather models are built to make forecasts and projections that are useful for understanding the behavior of the climate system and also useful to society and to stake-holders whose livelihoods or safety depend on reliable information about the future. However, the information sought by these stake-holders is not always directly produced from the models, and even less commonly is it used to evaluate the quality of the models. US CLIVAR works to develop linkages between communities so that the answers that decision-makers need are articulated during the model evaluation phase, and the models can be tested and diagnoses made or avoided if the models are deemed inadequate for the task. Section 6.5 discusses some of the ways these connections are made.

Just as standards of quality control of data and models are needed during model development, they are also needed during model evaluation. US CLIVAR seeks to ensure accurate methods for determining confidence limits, hypothesis testing procedures, and generally statistical methods capable of determining just when a model "failed" or "passed" an evaluation process. These statistical measures are particularly challenging for ensemble or probability forecasts. When an event occurs, such as El Niño or a drought, it is an excellent test of a model to see if the model foresaw the possibility of that occurrence. But how do we evaluate the probability the model associated with that outcome, as we never visited any of the other possibilities in the real world to see just how probable it was? Without such tools, model evaluations are prone to be too subjective to constitute an effective evaluation.

A new model is no better than an old model until it is proven so through evaluation against observations. The process involved in the evaluation of new generations of models as compared to the older generations is likely to be subjective and prone to confirmation bias. Confirmation bias is the tendency to point out successes and ignore failures of a beloved theory or model. The years spent by modelers and their collaborators to produce a new model makes it tempting to insist that the new model is better in nearly every regard over the old, or at least to down-play the faults of the new model. To avoid this psychological lapse, standards for comparison and evaluation need to be carefully developed and maintained over generations of models. These are especially important in assessing prediction skill of models, which can depend sensitively on the precise diagnostics and metrics chosen. Once a representative and revealing set is developed, additional effort is required to communicate the results of the improvements or failures to improve prediction skill to the user community who will use the model and its predictions operationally.

Establishing improved communication about models and their improvements is important. US CLIVAR has a number of regular and purpose-built communications (e.g., the newsletter US CLIVAR *Variations*), but their efficacy needs to be continually evaluated to ensure that they meet community needs. Would shorter, longer, less detailed, or more detailed articles improve their readership and transfer of knowledge to the community? Likewise, standards for model manuals, process study synthesis datasets, observational atlases, and scientific writing and reporting are all improved through the project reviews and communications from US CLIVAR.

4.5 Goal 5: Collaborate with research and operational communities that develop and use climate information

The climate is more than a physical system. Fluctuations and long-term trends in climate have far-reaching impacts on a wide variety of systems ranging from biological to social. In turn, the climate is sensitive to biogeochemical processes, including but not limited to those affecting the composition of the atmosphere, especially on multi-decadal and longer timescales. US CLIVAR is in a position to promote the kind of multi-disciplinary research required to make continued progress on the causes and effects of climate variability and change. This may be accomplished by taking advantage of US CLIVAR's core strengths related to the role of the ocean in the global climate system.

There are two types of multi-disciplinary problems involving the climate. One type features fundamentally interactive processes linking the elements. A prime example is represented by the carbon cycle and its relationships to the longer-term variability in the atmosphere-ocean system, which is an area of increasing CLIVAR collaboration. Many other phenomena do not involve two-way connections between disciplines, but involve assessing the importance of the climate forcing on the system of interest. A cogent summary of these linkages, organized by sector, is provided by the 3rd Report of the National Climate Assessment (McCarthy et al. 2001). A notable example is the identification of the Pacific Decadal Oscillation (PDO) that occurred as a result of collaboration between biologists and climatologists aimed at elucidating the variability in Pacific salmon catches (Mantua et al. 2010). This kind of multi-disciplinary discovery rarely results from physical scientists providing information on the past and future states of the climate alone—collaborations between scientists are needed to formulate the questions and bring the needed skills to bear on problems between disciplines.

Because of US CLIVAR's emphasis on the role of the ocean, the marine ecosystem is an example of multidisciplinary work where US CLIVAR can be effective. The research in this sphere has ranged from small, individual studies to large and long-lasting multi-agency programs such as the Global Ecosystem Dynamics Program (GLOBEC). These endeavors are continuing, for example, as part of the Integrated Marine Biogeochemistry and Ecosystem Research (IMBER) project. This type of work has produced new insights into the connections between the atmosphere-ice-ocean system and the biology. For example, high-latitude climate variability in association with the NAO is found to freshen the Labrador Sea water that flows downstream onto Georges Bank, which leads to earlier spring phytoplankton blooms, and ultimately, higher survival rates for young haddock. Another example of the bottom-up effects of the physical environment on food webs is represented by a collection of studies on the delayed upwelling in the California Current System that occurred in the spring of 2005. One lesson learned is that some impacts of climate variability do not result from the most commonly studied and understood modes of variability such as ENSO. This illustrates that it is often the less-appreciated aspects of the climate forcing (e.g., regional expression, timing of seasonal changes relative to norms) that are key to the system of interest. The legacies of these programs are the tools and models developed under their auspices.

Observational collaborations

Observing networks are becoming more multi-purpose and integrated. This development is central to the Framework for Ocean

Observing recommended by an OceanObs'09 working group (Lindstrom et al. 2012). This recommendation specifically called for "integrating feasible new biogeochemical, ecosystem and physical observations while sustaining present observations, and considering how best to take advantage of existing structures." Certainly coincident physical and biogeochemical measurements are essential for documenting linkages on the process level, but there is more to it than that. Each measurement can help justify other types of measurements. Such a philosophy is behind the monitoring with suites of instruments (chemical as well as physical) at the Ocean Climate Stations and some NOAA Ocean Reference Stations, and the enhancement of ARGO floats with optical sensors to estimate backscatter and chlorophyll concentrations. Similar relationships pertain to a whole host of other components of the Earth system, including terrestrial ecosystems, agriculture, and water resources among many.

The processes linking the physical to biogeochemical components are not fully understood in general, and so there is a continuing need for process studies. Multi-disciplinary field operations often involve the need for compromises. Nevertheless, these kinds of campaigns often provide a rationalization for detailed physical measurements. Using the marine ecosystem again as an example, the integrated ecosystem research programs carried out in Alaskan waters with the support of the North Pacific Research Board (NPRB), NSF, and NOAA are revealing physical phenomena and insights important in their own right. US CLIVAR has reviewed biogeochemical process studies, and supports the building the collaborations between the physical and biogeochemical communities through working groups and climate process teams.

Data interoperability is an issue of special importance to multi-disciplinary research. Different disciplines have different protocols and analysis tools; to a certain extent this encourages a variety of approaches in a positive way. However, the different approaches and protocols often are an impediment to multidisciplinary collaboration. Effective data management is always the goal, but for cross-disciplinary studies it is imperative. US CLIVAR has a history of helping evaluate and advise observational efforts (monitoring and major process studies) focusing on physics, and there will probably be opportunities to expand this role into interdisciplinary realms (e.g., through GEWEX, OCB).

Fostering collaboration between climate experts and regional experts is a key step in assessing regional impacts of climate variability. US CLIVAR often focuses on climate timescales and global variability. And while much of US CLIVAR focuses on regional processes that affect these larger, slower scales, US CLIVAR also poses the fundamental science question (Chapter 3.4), "What determines regional expressions of climate variability and change?" Working in collaboration with experts on collecting regional observations and understanding climate variability impacts on regional scales is a direct way for US CLIVAR to address this fundamental science question.

US CLIVAR has decades of experience in helping to design, and evaluate, and restructure complex observations and monitoring systems. Many of the US CLIVAR skills, best practices, and evaluation procedures will benefit multi-disciplinary and integrated observing systems as they have done for the physical observations of the past.

Modeling collaborations

The models that have been developed as part of multi-disciplinary research are of three basic types: (1) conceptual/qualitative, (2) statistical/empirical, and (3) dynamical. The distillation that occurs in the production of conceptual models is arguably especially useful for multi-disciplinary problems. These problems often involve systems that are complex, and hence benefit from simplification, as long as essential interactions are included. More quantitative statistical and dynamical models each have their benefits and drawbacks. The former are well-suited for situations with long time series for model training and testing, but are generally ill-suited for characterizing interactions involving functional relationships that are time dependent. This is less of a constraint for dynamical models, in principle, but these rarely include all of the possible processes influencing a system, and hence also may not be able to account for all the different ways it can evolve. It is worth mentioning that many of the interactions in dynamical models linking the physics to other aspects, e.g., the biology, are based on functional relationships fitted to experimental results, with their attendant errors and uncertainties, rather than fundamental principles such as those encapsulated in the Navier-Stokes equations—albeit fundamental principles that were originally discovered through a process of fitting to experimental results.

The challenges in modeling complex systems suggest the utility of multiple approaches: US CLIVAR would encourage

development, evaluation, and comparison of different type (statistical, dynamical and hybrid) models. This kind of effort has been carried out for ENSO models; the degree to which different kinds of models agree is itself valuable information. These sorts of model intercomparisons should be done in both hindcast and forecast mode. These tests should be structured such that it can be determined how much skill is limited by errors in the physical climate forcing itself versus the errors associated with simulating the mechanistic effects of the forcing on the properties of interest.

US CLIVAR aims to aid in model development of both climate/Earth system models and operational models used for weather forecasting, which serves to foster coordination and collaboration between the communities that develop these models. Too often the tools and methods of these communities have been developed in parallel with limited communication. Consequently, algorithms, code, data formats, model evaluation metrics, and the results of experiments are often rediscovered, reinvented, or are lost to some communities. Evaluation techniques developed in the operational community can find important uses in the climate community and vice versa. The concept of "seamless" forecasting and model development can foster a variety of modeling techniques and foci so that models span many possibilities. US CLIVAR can foster better coordination through joint workshops, directed research initiatives, and best practices guidelines so that this duplication of effort can be reduced to mutual gain.

Communication and collaboration must occur between communities of modelers. US CLIVAR can aid in this communication and collaboration by development and promotion of standardized data formats (such as NetCDF), so that model results can easily be shared. Distribution and formatting of observational data from existing and new observing systems is also often a substantial barrier to their incorporation into models and use as tools in model evaluation. Collaborations between CLIVAR and CliC on ocean-ice modeling, between CLIVAR and the Carbon Cycle modeling community, as well as CLIVAR and weather forecasting communities, are good examples of potential collaborations to improve modeling efforts.

Programmatic subgoals

To effectively collaborate, working relationships across disciplines are necessary. It simply takes time together to begin speaking in a common language and to become familiar with potential collaborators' interests, resources, and limitations. Although this occurs in an ad-hoc manner, workshops offer a focused forum for larger groups working on interdisciplinary problems. There exist plenty of examples of successful multi-disciplinary programs (e.g., GLOBEC) from which lessons may be learned. Ideally, it should be possible for researchers to receive support from different core programs and via targeted calls for specific research. US CLIVAR could help with this process by seeking greater participation of scientists with experience working with colleagues from other disciplines. Those types of scientists would then be in a position to advise on the type of climate research needed for different sectors, and to serve as messengers on the latest in climate research. Early-career multidisciplinary training, of postdoctoral scientists and students in particular, is a key way to broaden the expertise of those scientists and improve the training and communication of the community as a whole. Section 6.5 explains further the methods that US CLIVAR uses to encourage these interdisciplinary connections.

Working groups have traditionally studied the physics of the climate system, but now US CLIVAR seeks to address multidisciplinary topics such as the two working groups focused on marine carbon established in 2012. Additional US CLIVAR working groups may be particularly effective in terms of synthesis. Even though synthesis is always a goal, the individual pieces of the research almost inevitably get the majority of attention, and funding is often short before all the pieces are fully assembled. US CLIVAR working groups can take advantage of the benefits of hindsight as well as multiple perspectives.

The effects of past and present climate variations on other systems are of continuing interest. Improved understanding of these linkages is necessary to better anticipate and mitigate the probable impacts of climate change. US CLIVAR is currently playing a meaningful role in this overall effort on the specific topic of marine carbon, and other opportunities are liable to present themselves. This contribution could be in the form of direct involvement with other specific topics and perhaps also as a clearinghouse for climate information for wider audiences.

The interdisciplinary facet of US CLIVAR may provide information that is particularly relevant to resource management. Hence it would be preferable to include the participation of experts that have experience translating research results into forms suitable for stakeholders and policy makers, such as those being

trained through PACE (Chapter 2). This can only help broaden US CLIVAR's audience and impact.

4.6 Conclusions

The goals presented in this chapter expand upon the US CLIVAR mission and will guide the activities for US CLIVAR over the coming years. Addressing and achieving these goals will be a challenge for all of the scientists and agencies associated with US CLIVAR. Successfully doing so will benefit society and the Earth sciences, deepening our understanding of our environment and how to sustainably fulfill our role in it. Upcoming chapters will describe some of the research challenges that presently are being attacked to address these goals as well as cross-cutting strategies that US CLIVAR will use to help scientists and agencies coordinate to achieve these goals.

Chapter 5
Research Challenges

US CLIVAR supports all activities that advance its Mission (Chapter 1), address its fundamental science questions (Chapter 3), and support its Goals (Chapter 4). In addition to continuing science activities related to those it has carried out in the past (Chapter 2), US CLIVAR will also continue to highlight a small number of Research Challenges. Research Challenges are broad areas of climate science that are societally important, reflect the interests of the scientific community and funding agencies, concern most (if not all) the US CLIVAR panels, and typically extend US CLIVAR beyond its traditional research agenda. They are initiated only after thorough discussion by the US CLIVAR community at a Summit conference. Because of their breadth, they are expected to last throughout much of the program.

Ongoing Challenges concern decadal variability, climate extremes, polar climate, and interactions between climate and carbon/biogeochemistry. This chapter provides definitions of each of them, explains how the research is linked to the ocean, discusses why the Research Challenge is of interest from both a scientific and societal point of view, examines what is known about the underlying dynamics, and suggests what future research is needed to further our understanding of the causes and improve predictability. It is expected that additional Research Challenges will be added to the US CLIVAR agenda as the need for them arises.

5.1 Decadal variability and predictability

Climate exhibits variability on decadal (10–20 year) timescales, with important societal consequences. Decadal climate variability is often large enough to overshadow regional and global anthropogenic trends, and hence has relevance for guiding planning decisions about future adaptation investments. At the same time, developing a comprehensive dynamical understanding of decadal variability and developing predictive skill based on decadal modes has proven difficult, primarily as a consequence of the scarcity of *in situ* time series that are long enough to resolve it with statistical reliability. Ongoing challenges are to identify causes and determine if they can be exploited for decadal climate prediction, and to separate natural from anthropogenically-forced variability in the evolution of the climate system (Solomon et al. 2011).

Definition

Decadal variability is generally described in terms of large-scale modes (Deser et al. 2010). These modes are the dominant patterns of variability, involving both atmospheric and oceanic variables, which reoccur (oscillate) at decadal timescales and are concentrated in specific regions. Several modes can exist within a region. In that case, they tend not to be completely independent, either because they share common dynamics or because similar variables are involved in their definitions. For example, several of the modes describing decadal variability in the Pacific Ocean region are correlated with each other to some degree.

In the Pacific region, one example is the Pacific Decadal Oscillation (PDO; Mantua et al. 1997; Liu 2012). Figure 5.1 illustrates the PDO, showing the spatial pattern and time dependence of its warm and cold phases. The spatial extent of the PDO is striking, as it extends throughout much of the basin and also affects climate variability in North America. Other Pacific decadal modes include: ENSO decadal variability (EDV; e.g., Graham 1994; Newman 2007; Di Lorenzo et al. 2010; Furtado et al.

2012), which happens because ENSO differs significantly from one event to another; the Interdecadal Pacific Oscillation (IPO); and the North Pacific Gyre Oscillation (NPGO; Di Lorenzo et al. 2008). Modes of Atlantic decadal variability include the Atlantic Multidecadal Oscillation (AMO; Enfield et al. 2001), North Atlantic Oscillation (NAO; Hurrell 1995), and Tropical Atlantic Meridional Mode (TAMM). The Indian Ocean does not exhibit a distinct mode of decadal variability (Lee 2004; Lee and McPhaden 2008). Modes of low-frequency variability in the polar regions include the Arctic Oscillation (AO, also known as the Northern Annular Mode) and the Antarctic Oscillation (AAO, also known as the Southern Annular Mode).

The time dependence of climate modes typically exhibits a broad peak at decadal frequencies. The lack of a distinct peak is certainly one reason why isolating the basic processes that underlie decadal variability has proven difficult. Decadal-scale variability may also be weak relative to the year-to-year variability. A relatively low signal suggests that the predictability associated with decadal variability is limited (but see below).

Significance

Decadal variability is societally important because it directly impacts atmospheric, terrestrial, as well as oceanic conditions. For example, the PDO influences Australian rainfall as much as ENSO (Power et al. 1999a,b; Meinke et al. 2005; McKeon et al. 2009), TAV impacts rainfall in the Nordeste region of Brazil and the Sahel (Zhang and Delworth 2006), and there are decadal variations in the Asian monsoons. Decadal variability of river flow into reservoirs is important for water supply and hydro-power planning, with decadal (and longer) periods of above- and below-average flow determining system yield more than long-term annual averages (Bureau of Reclamation 2007).

Another important consequence of decadal variability is that it obfuscates the climatic trend expected from increasing greenhouse gases, causing global-mean temperatures to rise more rapidly or slowly in some decades (Solomon et al. 2011). For example, the rapidly increasing global mean temperature rise since the 1970s has slowed or even halted since the 1990s (Easterling and Wehner, 2009; Kaufmann et al. 2011, Meehl et al. 2011) and wintertime northern-hemisphere temperatures have cooled during the last 10 years (Cohen et al. 2012a,b). At least for some variables (e.g., precipitation), the trend is not expected to emerge as the dominant climate signal until the middle of the century (Deser et al. 2012b). In the interim, separation of natural

(decadal) variability from the anthropogenic signal will be difficult, potentially leading either to costly adaptive strategies or to a sentiment of complacency to the need for such actions.

Dynamics

Despite extensive research, the mechanisms that lead to decadal climate variability are not yet well understood. Because of the large thermal inertia of the ocean, it is likely that decadal variability has its origins in coupled ocean-atmosphere interactions, as is known to be the case for the prominent modes of interannual variability (e.g., ENSO in the Pacific, Atlantic Niño, and the Indian Ocean Dipole). In addition, other factors that are not directly linked to ocean-atmosphere coupling are known to impact regional climate variability, and may also be important for decadal variations and feedbacks. They include soil moisture in deeper layers, vegetation, snow cover, changes in anthropogenic aerosols (Kaufmann et al. 2011, Booth et al. 2012), and stratospheric water changes (Solomon et al. 2011).

To illustrate, consider decadal variability in the Pacific Ocean. One hypothesis for its cause is midlatitude SST variability generated by stochastic forcing, in which random atmospheric surface forcing with a "white noise" spectrum is integrated by the ocean mixed layer to produce "red noise" (Newman 2007). This process alone, however, cannot account for spectral enhancement in the decadal band. Processes that have been proposed to account for such enhancement include: changes in the North Pacific oceanic gyres and the latitude of the Kuroshio Extension (Di Lorenzo et al. 2008); fluctuations in the strength or water-mass properties of the shallow overturning cells (Schott et al. 2004) that connect the subtropics to the tropics (Gu and Philander 1997; Kleeman et al. 1999; Nonaka et al. 2002; Solomon et al. 2003); tropical-to-midlatitude teleconnection of ENSO decadal variability (Newman 2007); and surfacing of temperature anomalies created in prior years due to an anomalously strong, seasonal thickening of the oceanic mixed layer. A similar range of physical mechanisms has been proposed for Atlantic decadal and multi-decadal variability (Kaufmann et al. 2011; Booth et al. 2012). For example, Booth et al. (2012) argued that the indirect effect of aerosols is a prime driver of the observed Atlantic multi-decadal variability. On the other hand, Zhang et al. (2013) showed that the model simulations considered in Booth et al. (2012) have major discrepancies with observations in large part caused by aerosol effects, which casts considerable doubt on the Booth et al. (2012) conclusions. Quantifying the relative importance of ocean circulation and aerosol forcing on Atlantic multi-decadal variability remains an important challenge.

Pacific Decadal Oscillation

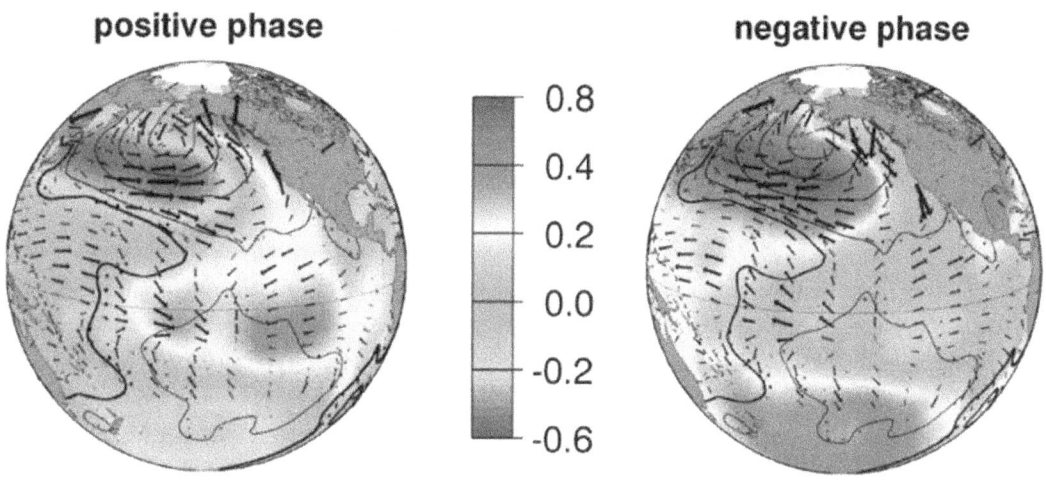

positive phase   negative phase

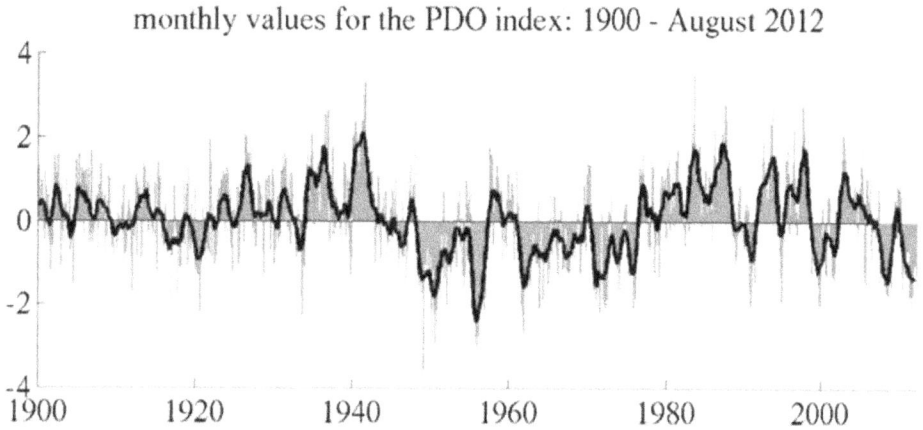

monthly values for the PDO index: 1900 - August 2012

Figure 5.1

(top) *Wintertime anomaly patterns of SST* (color shading; °C), *sea-level pressure* (contours; hPa), *and surface wind stress* (arrows; longest values ~0.015 N m⁻²) *associated with warm* (left) *and cold* (right) *phases of the PDO.* (bottom) *Index of the PDO, defined by the leading principal component* (PC) *of monthly SST anomalies in the Pacific Ocean north of 20°N for the 1900–2012 period. The index is normalized so that its unit corresponds to a standard deviation of the data. The solid black line is a 3-year running mean of the index.*

Source: JISAO, University of Washington

Future research

Many questions remain about decadal variability: its character, the processes that generate it, the scope of its predictability, and hence the level of predictive skill. Scarcity of observations is a severe stumbling block in answering these questions. The use of coupled models therefore becomes a critical tool for scientific advancement. Property-conserving state estimation (data assimilation) is also useful, helping to extract as much information as possible from scarce datasets and to facilitate the improvement of process parameterizations in coupled systems. It is essential to continually evaluate and improve the fidelity of these systems (e.g., Collins et al. 2006; Furtado et al. 2011).

In addition to their ability to improve dynamical understanding, coupled models are beginning to be used to explore the predictability of decadal variability, and experimental prediction efforts have begun (Smith et al. 2013). Some early studies have shown potential decadal predictability when oceanic decadal anomalies can be tracked back to a specific oceanic source, typically a subsurface record indicative of past air-sea interactions. For example, changes in the AMO are associated with AMOC variability (Knight et al. 2005; Zhang, 2008, Hawkins et al. 2011), and the latter has decadal predictability (Teng et al. 2011). On the other hand, the models where these signals are predictable are simple in comparison to the real world, in which strong higher-frequency variability makes predictions at decadal timescales more challenging.

Many outstanding practical issues must be addressed before decadal predictability can be utilized in coupled models to make predictions. They include the following. Given imperfect and incomplete observations and assimilation systems, what is the best method of initialization? What is the added skill in climate predictions with initialization when compared to uninitialized predictions? What is the impact of small ensemble size in the spectrum of decadal means? What predictions should be attempted, and how would they be verified?

5.2 Climate and extreme events

Extreme meteorological and oceanographic events can have major impacts on society. Their properties (e.g., strength, duration, and frequency) are often clearly linked to large-scale climate variability and change. In many cases, however, the linkages are so complex that they are not yet adequately understood or even their existence is a matter of debate.

There are many examples of such extremes. Meteorological examples can be grouped into categories related to: temperature, precipitation, wind, sea level, thunderstorms, and larger scale weather-systems —like tropical and extratropical cyclones. Oceanographic examples include wave height, coastal inundations, ice surges, anoxic regions (dead zones), coral bleaching, and red tides. This section discusses general properties of extremes and their relation to climate, focusing on four general categories: extremes associated with hurricanes, precipitation, droughts, and temperature.

Definition

Extreme events are generally defined in terms of events that occur relatively rarely (e.g., seasonal precipitation totals that occur less than one percent of the time) or that exceed some threshold (e.g., temperatures greater than 90°F in non-arid regions). Due to the very wide range of possible definitions, the scope is rather vast. For example, extremes can be defined for any variable and for any timescale (e.g., hourly, daily, weekly, and seasonal-to-decadal for precipitation extremes). Additionally, their magnitude is often defined in alternate ways, for example, either as an absolute measure (e.g., a 10-cm threshold for precipitation) or a ranking (e.g., the upper 1% of daily rainfall events). The underlying physical processes and societal implications can vary considerably based on these definitions, even for the same variable and timescale.

The most often-mentioned climate-related extremes are: hurricanes and other intense-wind events; heavy-wave and storm surge events; heavy precipitation on timescales from hours to days, which can lead to flooding on timescales from days to weeks; droughts on timescales from weeks to decades; and cold snaps and heat waves on timescales from hours to weeks. It is generally possible to distinguish between extremes happening on timescales from days to several weeks—heavy precipitation, heat waves, cold snaps, high winds—and extremes occurring over seasons or longer—primarily droughts. Some events, however, blur this distinction, such as season-long flooding episodes and heat waves. Moreover, as precipitation is always individually produced from short-term processes (hours to days), the important dynamics must, at some level, always include the shorter timescales.

The ocean impacts climate extremes primarily through its impacts on cyclogenesis and the large-scale atmospheric circulation. In this regard, the various climate modes have a strong

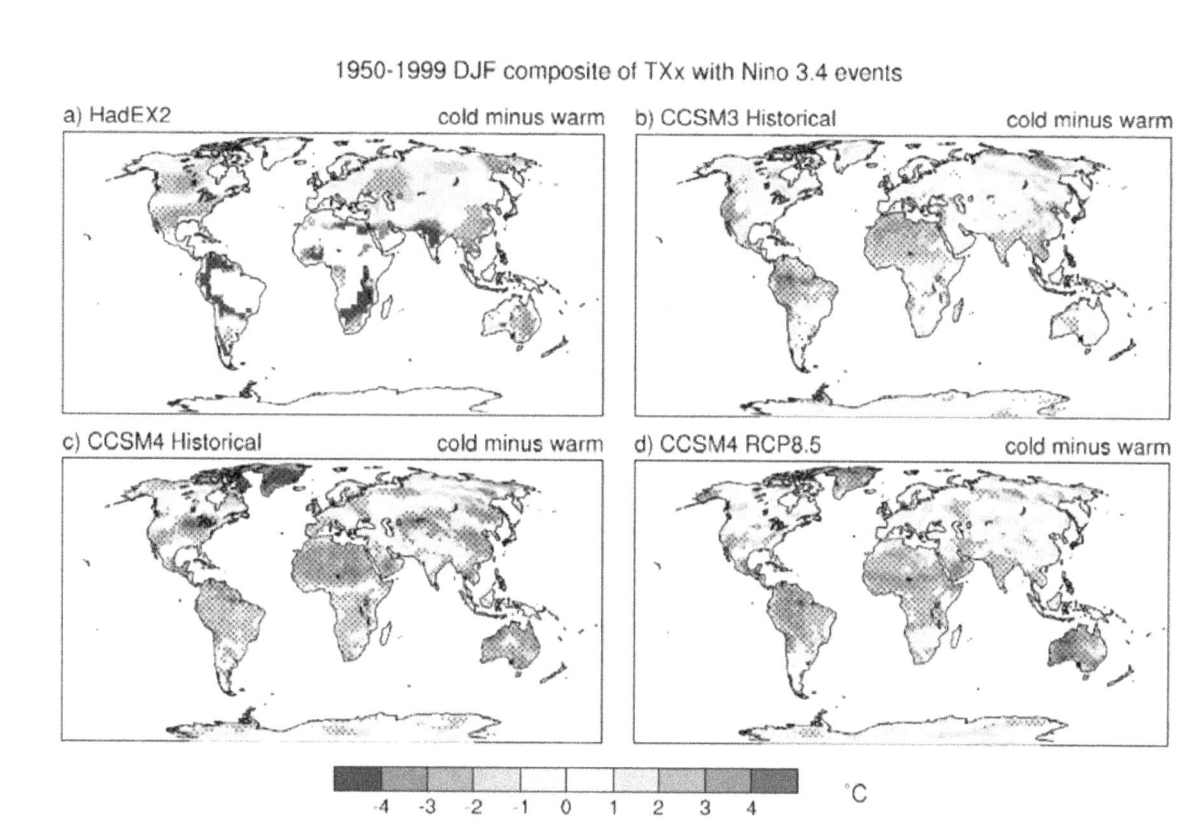

Figure 5.2

Composites of maximum daily temperatures for cold minus warm ENSO events during DJF from four different climate-model simulations: HadEX2 (top left), CCSM3 Historical (top right), CCSM4 Historical (bottom left), and CCSM4 RCP8.5 (bottom right). For example, maximum temperatures are warmer over much of the US (warm colors) during strong El Niño than during strong La Niña events. Stippled regions pass a 5% t-test. White areas in a) denote grid points with insufficient spatial coverage. See Arblaster and Alexander (2012) for details and other references.

influence on modulating extremes, although this relationship has been better explored for some types of extremes (e.g., precipitation and drought) than for others. For example, climate modes can shift preferred locations of convective precipitation (e.g., ENSO and the MJO) that in turn influence the midlatitude circulation (e.g., PNA and cold air outbreaks, ENSO and summer maximum temperatures). They are linked to droughts and winter maximum and minimum temperatures extremes over North America (e.g., to the PDO), determine whether precipitating events even exist (e.g., the impact of Indian Ocean Dipole events on MJOs), and influence the severity of events (e.g., the

impact of SST on tropical cyclones). Figure 5.2, for example, illustrates impacts of ENSO on global land temperatures.

Significance

Scientific significance

Extremes are interesting scientifically from several viewpoints: *i*) extremes are, by definition, rare, unusually strong, or both; *ii*) they involve a complicated mix of timescales and processes; and *iii*) they provide a stringent test of model capabilities in reproducing complicated dynamics. The rarity of extremes makes them challenging to observe and characterize; at the same time, however, their unusual intensity makes them compelling and perhaps "easier" for study, in that they stand out strongly from background conditions. Extremes are a particularly compelling bridge between weather scales and climate scales. They are short-term events with long-term consequences. The significance of an extreme can be amplified when combined with another variable, such as high humidity during an episode of high temperature. The possibility that pending climate change could make extreme events more commonplace or more intense makes the understanding and prediction of extreme events of critical importance.

Societal importance

Extreme events are societally important because they often have devastating societal and economic consequences. From 1980 through 2011, there were 133 weather-related disasters in the US with damages that exceeded one billion dollars (adjusted to 2012 dollars), with annual loss totals as high as 187 billion dollars (NCDC 2012, Smith and Katz 2013). By their very nature, some extreme events such as hurricanes, heat waves, and cold snaps are short-lived and rare, but they can have disproportionately large societal impacts. For example, heat waves cause a larger annual number of deaths (170) than either hurricanes (117) or flooding (74) from 1997–2006, and there were 500–1000 fatalities attributed to the Chicago heat wave of 12–15 July 1995. Cold air outbreaks can carry large economic losses. The timing of the outbreak can be more important than the minimum temperature of the freeze. For example, during 4–10 April 2007, low temperatures across the southern US caused two billion dollars in agricultural losses, since many crops were in bloom or had frost sensitive buds or nascent fruit. During December 1989, two days of subfreezing weather wiped out half the citrus trees in Florida, even though the monthly mean temperature was above normal. Hence, monthly means can be misleading as several types of extremes with high impact last for such a short time that they do not necessarily appear in monthly mean data. Many more examples and extensive discussions of extreme events are found in IPCC SREX (2012).

Dynamics

Within the broad category of climate and extreme events, impacts of climate on droughts and hurricanes have seen the most general study. Other categories have typically been addressed more in terms of regional or case studies.

Droughts

Regarding droughts, forcing by Pacific and Atlantic SST anomalies associated with climatic modes (e.g., ENSO and the PDO) appears to have played a prominent role in most major US drought episodes, with additional influence from local factors (soil moisture, temperature-driven evaporation, water availability, vegetation cover and state, etc.). While connections to SSTs in both observations and modeling studies are fairly robust, capturing the magnitude of severe droughts remains difficult. Whether errors result from random noise or imperfect representations of the underlying dynamics is not yet clear. Furthermore, the specific mechanisms by which the large-scale circulation anomalies associated with oceanic forcing modulate continental precipitation are still a current research question.

Hurricanes

Hurricane dynamics have been the subject of considerable study, and existing models are able to reproduce many aspects of their movement and distribution (strength, pathway, and existence, etc.). These properties are known to depend critically on oceanic conditions, not only SST but also near-surface heat content. As such, hurricane properties (e.g., their frequency and intensity) are closely linked to large-scale climatic variability, such as El Niño events, the NAO, and the AMO. How their properties may change under global warming is less certain, depending on multiple factors that can either enhance or suppress cyclogenesis. A few studies have investigated the contribution of hurricanes to the occurrence of extreme precipitation over the US, but the dynamical factors that determine the magnitude of precipitation during hurricanes are not yet well understood. Additionally, hurricanes are related to the occurrence of extreme values of winds and storm surge, which are important extremes in terms of societal impact.

Precipitation

The climatic factors influencing precipitation extremes are poorly known, in part due to the broad range of dynamical mechanisms involved and to regional differences. Precipitation extremes are known to occur in association with synoptic storms, tropical cyclones, and organized heavy convection; atmospheric rivers are known to be an important factor on the US West Coast. While some extreme events can be very local, synoptic and regional factors impact them and may provide the basis for statistical correction of model output and predictions. A link has been shown between short-term precipitation extremes and some modes of ocean-related variability including ENSO and Pacific decadal variability, suggesting the possibility of long-lead prediction. An overall assessment of the key causes of, and large-scale influences on, extreme short-term precipitation for the United States, however, has not yet been made.

Temperature

The causes of temperature extremes (heat waves and cold outbreaks) are not fully understood. Heat waves and cold-air outbreaks are associated with large displacements of air masses into regions where they are not normally found, which in turn are caused by unusually large meridional oscillations in the circulation. Factors that influence heat waves include both local and remote larger-scale factors. Large-scale circulations that create topographic slope flows (adiabatic warming by sinking) or flows that block moderating temperatures (e.g., creating an offshore pressure gradient that opposes cooling sea breeze in California) can amplify temperatures locally. Soil moisture content, vegetation type, and irrigation are local factors that impact the intensity of hot spells. Accordingly, hot spells are exacerbated by drought conditions; conversely, the high temperatures can amplify the impact of drought. Climate models are able to generate large-scale patterns with extreme heat (e.g., Meehl and Tebaldi 2004), although important details of the large-scale pattern as well as important local processes may be missing; further, the amount of variability may be incorrect or correct for the wrong reasons (Grotjahn 2013).

Future research

The significant gaps in our basic understanding of the causes of climate extremes are limiting our ability to make physically-based predictions and projections. Two key areas where even our basic dynamic understanding is limited are: the dynamics of short-term precipitation extremes not associated with tropical cyclones, and the dynamics of short-term temperature extremes (both heat waves and cold snaps). Key questions are: What are the important dynamical processes that underlie short-term precipitation and temperature extremes? How do these short-term processes interact with the larger-scale, slower and potentially-predictable climate fluctuations linked to the ocean? What are the timescales, metrics, statistics, and analysis tools that are most relevant for extremes, both to their dynamics and societal impacts? What properties of extremes, if any, are changing under global warming? Additionally, although there have been some recent efforts to codify a set of extreme definitions such as by the joint CCl/CLIVAR/JCOMM Expert Team on Climate Change Detection and Indices (ETCCDI), some community-level discussions on terminology and language would be helpful.

Observational issues

Having an adequate observational database is essential for understanding extremes. In this regard, it is important to ensure that records are long lasting and without gaps in order to capture reliable statistics, especially if the statistics are not stationary. It is also important that records include all relevant variables, since for some extreme events combinations of variables can be deadly, even though the individual values of the combination are not so extreme (e.g., elevated humidity with elevated temperatures; cold temperatures with strong winds).

Modeling issues

Regarding modeling, key questions are: Do current models produce realistic precipitation and temperature extremes for the correct dynamical reasons? Do they adequately represent both large-scale and local processes? Are model parameterizations, which might be tuned for "median conditions," adequate for representing extremes? Are they therefore reliable enough to make long-term projections? Answering these questions is critical, given that inferring trends in extreme events is a regular part of the IPCC and US global change reports. In many cases, the answers are unknown or appear to be no. For example, while models can approximate the large-scale environment associated with heat waves, important properties of the near-surface temperature field are lacking in an absolute sense, if not also in a relative sense (accounting for systematic model bias). Part of this lack is certainly due to model resolution, which is often inadequate to resolve complex topography. Another part is likely due to inadequate model parameterizations (e.g., soil moisture). A third may be that the large-scale circulation is not sufficiently well represented. For example, Scaife et al. (2011) show how bias-correcting errors in a small region of North Atlantic SST have a large impact on winter blocking frequency (i.e., cold air outbreaks are linked to some blocking patterns in models).

Infrastructure

Modeling of climate extremes will benefit from advances in raw computing power, as high-resolution models are needed to represent features at regional and smaller scales. Output from high-resolution model runs is valuable for diagnostic studies, regional climate simulations, and potentially for application-sector users. Currently, however, nearly all of this output is discarded due to the large volume that is created. Some daily output from CMIP5 is archived, but only for a limited set of fields, and the outputs are fairly difficult to access. Infrastructure improvements and new technologies are needed to make this output more accessible to users, reduce storage costs, and facilitate analysis.

In addition to more computing power, infrastructure is needed to understand and to characterize extreme events better. While simple indices (e.g., CLIVAR's ETCCDI extreme indices) are a useful start, further improvements are needed such as: application of extreme value statistical methodology and application of advanced dynamical tools to gain a physical understanding of the extreme phenomena. Extreme value statistics has its own tools and procedures requiring specific expertise in order to use these properly. The proper statistical characterization of extreme events properties and a dynamical understanding of the extreme events are needed, especially in the context of a non-stationary climate. Both extreme statistics and advanced dynamics are needed to address key uncertainties in extreme event probabilities, from simple return period changes to attribution studies.

Applications

Model results need to be more suitable for various application sectors (water managers, transportation, energy, agriculture, public works, insurance/reinsurance, forest managers, etc.). The information needed to assess the impacts of high or low temperatures on agriculture is multivariate (temperature, humidity, winds in combination) and needs to be at a sufficiently high enough time frequency (hourly). Other sectors (e.g., water managers) need different time and space resolution as well as other quantities of interest. Most application sectors need surface data that is less well simulated than upper-air variables. It may be unreasonable to seek model output at the needed high resolution but that leads to two general points: first, quantifying the associated uncertainty is useful to application sectors as these workers are often familiar with the concept of uncertainty; and second, there may be a general need for providing and evaluating model output of the type needed by application sectors, including the research communities described in Section 4.5. Interest in this is recently growing in the modeling community, for example, the latest working group for NCAR's CESM program.

Comprehensive research programs

Given their complexity, extremes research will benefit from, and indeed requires, targeted, systematic research programs that include both observations and modeling. Such programs need to include both process and regional studies, with a common overall effort that guides the approach to definitions and methodology and which would allow the results to be easily compared and summarized. Previous efforts that focused on droughts and

hurricanes have proven very effective. They should be supplemented by additional focused efforts (e.g., short-term precipitation extremes, short-term temperature extremes, integrated analysis of the utility of metrics for extremes).

5.3 Polar climate changes

Climate-change signals are amplified in polar regions, and their manifestation there (e.g., the disintegration of ice shelves and glacial melting) has helped to make the public aware of the consequences of a warming world. Polar climate change can also have a profound influence at lower latitudes (e.g., sea level rise through ice sheet melting, changes in global ocean circulation). Despite their importance, polar regions are inadequately observed, a consequence of the logistical challenges of data collection there. This scarcity has hampered our ability to understand, model, and predict the influence of polar climate change on the overall Earth system. For these reasons, observing and understanding polar climate change is a core US CLIVAR Research Challenge, already with several working groups dedicated toward the effort.

Definition

The polar regions play a fundamental role in the earth's climate system both by affecting the energy balance (they are regions of net heat loss, in contrast to the tropics that are regions of net heat gain) and through storage (or release) of large amounts of freshwater that can impact sea level rise and the global ocean circulation. Transferring the excess heat from the tropics to the poles through oceanic and atmospheric processes drives virtually all climate dynamics. At the same time, the two poles are quite different in their geography, climate, and interactions with lower latitudes. The Arctic Ocean is essentially a very large estuary in which stratification is largely determined by salinity, not temperature. Relatively fresh cold water sits above warmer saltier water below and deep mixing is limited. Exchanges with the world oceans are restricted to a few narrow straits through the continents that surround the Arctic Ocean. In contrast, the large Antarctic land mass is surrounded by the Southern Ocean, which facilitates interchange between the world oceans and where deep mixing is common. North of Antarctica, land coverage is relatively small and the oceans dominate climate interactions.

As a consequence of the different geographical configurations, sea ice cover in the two hemispheres has quite different characteristics and seasonal cycles. There is a larger extent of multiyear sea ice in the Arctic Ocean, whereas in Antarctica the dominant sea ice is seasonal with only limited areas of multiyear ice. In the Arctic, sea ice grows on the bottom, while in Antarctica the higher snowfall rates and larger snow accumulation sinks the ice so that much of the ice grows on the top due to flooding at the same time that it is melting on the bottom. These distinctions are important for understanding the different responses of sea ice at the two poles to anthropogenic warming. For example, how precipitation changes in a warmed climate will have a critical impact on Antarctic sea ice but not so much in the Arctic. Similarly, the different geographical configurations contribute to major differences in the two polar ice sheets: since all of the southern polar region is land-covered, the Antarctic Ice Sheet is ten times larger (by volume) than the Greenland Ice Sheet.

The polar regions play a crucial role in air-sea heat and gas exchange, water mass formation and transformation, and in the global oceanic circulation. In particular, south of 30°S the Southern Ocean occupies just under one-third of the surface ocean area, yet accounts for a disproportionate share of the vertical exchange of properties between the deep and surface waters of the ocean and between the surface ocean and the atmosphere. Models and observations indicate that: i) the Southern Ocean may account for up to half of the annual oceanic uptake of anthropogenic carbon dioxide from the atmosphere (Gruber et al. 2009) and for most of the excess heat that is transferred from the atmosphere into the ocean; ii) the uptake of heat and carbon dioxide by the Southern Ocean will diminish due to reduced vertical mixing resulting from increased stratification and due to an increased upwelling of carbon rich deep waters (LeQuéré et al. 2007); iii) Southern Ocean winds strongly affect the AMOC (Toggweiler and Samuels 1998); and iv) the depth of Southern Ocean isopycnals affects the global stratification and overturning circulation (Wolfe and Cessi 2010; Kamenkovich and Radko 2011). At the same time, adequate simulation of the Southern Ocean remains one of the main challenges facing IPCC-type climate models (Russell et al. 2006). Models indicate that future changes in clouds over the Southern Ocean are critical for setting the Earth's climate sensitivity (Trenberth and Fasullo 2010) and that such changes can impact the location of the ITCZ (Hwang and Frierson 2013).

Significance

Changes in the polar climates impact a number of societal groups. In the Arctic region, these groups include people

interested in resources, such as petroleum extraction or fisheries, people who live in the region who are interested in the changing patterns of transportation and subsistence hunting, and people interested in ecosystem changes and wildlife management. While the southern polar region is less inhabited, climate change in this remote region can have significant impacts on the surrounding Southern-Hemisphere nations. For example, the increase in ultraviolet-B radiation associated with ozone depletion above Antarctica has limited marine phytoplankton productivity in the Southern Ocean (Karentz and Bosch 2001), as well as being linked to increased incidences in skin cancers in Australia (Lemus-Deschamps and Makin 2012).

The societal importance of understanding the response of the ice sheets to polar climate change is difficult to overestimate: melting of all present-day ice sheets have a potential to raise sea level by ~65 m (Lemke et al., 2007), and globally ~150 million people live in coastal areas within 1 m of present-day sea level (Rowley et al., 2007). The combined contribution of the Greenland, East and West Antarctic Ice Sheets to sea level exceeded 10 mm in the past 20 years (Shepard et al. 2012), and the rate of their contribution has been increasing in time (Rignot et al. 2011). Spatial and temporal patterns of the observed changes are highly non-uniform, and in many instances have different signs. Sizable mass loss has occurred at the margins of Greenland and West Antarctic ice sheets. While the exact chain of events that resulted in the mass loss is still being debated, an increasing number of studies suggest a warming ocean is the trigger (e.g., Joughin et al. 2012). In addition, the paleo-climate community is also using models to help understand natural ice-shelf volatility in response to oceanic warming, and to help provide a baseline as to how fast ice margins have prehistorically retreated without anthropogenic forcing. Nonetheless, to observe these changes in polar ice sheets, ice volume and extent happening at rates as fast or faster than predicted by models has captured public attention, and provides stimulus for further research on the representation of polar processes in the models and in the relative roles of natural versus anthropogenic variability in enhancing (or limiting) the decline.

There are also indications that changes in sea ice coverage and enhanced warming in high northern latitudes, relative to the rest of the northern hemisphere, is impacting midlatitude weather and weather extremes (Francis and Vavrus 2012). The line of reasoning suggests that with enhanced warming in the far north, the equator-to-pole temperature gradient is reduced which in turn reduces the mean strength of the mid-latitude westerlies, enhances long-wave amplitudes, and slows the progression of long-wave structures. The latter increases the chance of blocking patterns and the persistence of dry or stormy conditions at some locations and hence the chance of extreme weather. However, the merit of this argument still warrants further research.

Dynamics

Significant changes have emerged in the Southern Hemisphere atmospheric circulation over the past few decades. An intensification and poleward shift of the subpolar westerly winds is thought related to the stratospheric ozone depletion (Thompson and Solomon 2002). Since these westerlies set the strength of the Antarctic Circumpolar Current (ACC) it is expected that this trend will also impact the dynamics and properties within the Southern Ocean, although there remains considerable debate as to how the ACC will respond. A steepening and poleward shift of isopycnals and fronts brought about by increased winds might be associated with an acceleration of the ACC, although this effect could be mitigated by an increased eddy field (Hallberg and Gnanadesikan 2006). Mesoscale eddies are thought to play a critical role in maintaining the Southern Ocean stratification (Marshall and Radko 2003). Atmospheric changes have already resulted in warmer and fresher water masses that are formed in the Southern Ocean (Gille 2008; Sprintall 2008; Purkey and Johnson 2010; Jacobs et al. 2011). Although consistent with an amplification of the global hydrological cycle (Durack and Wijffels 2010) and a warming atmosphere (Boning et al. 2008), these modifications could also be related to the poleward shifting of ACC frontal systems in response to wind changes. Changes in the wind field can also impact the uptake of carbon in the Southern Ocean, as strengthening winds act to intensify the upwelling of deeper carbon-rich water masses. Clearly, how the ACC dynamics and the carbon cycle might respond to these changes in the surface momentum and buoyancy fluxes are complex but pressing issues that are not easily unraveled.

In the northern polar region, the upper layers of the Arctic Ocean (and the Nordic Seas) are strongly connected to the North Atlantic Ocean through the import of warm, salty waters of Atlantic origin beneath the surface and the export of sea ice and glacial melt at the surface. There has been some evidence that the warming (Bersch et al. 2007) and slow down of the subpolar gyre that occurred in the mid-1990s (Hakkinen and Rhines 2004) are potentially affecting the Nordic Seas (Hatun et al. 2005) and the Arctic Ocean (Polyakov et al. 2004). In addition,

climate model predictions indicate a pronounced warming of the waters around Greenland (Yin et al. 2011).

The recent dramatic decrease in summer Arctic sea ice extent and thickness is perhaps the most visible consequence of the increase in greenhouse gases. The record low levels of ice extent during the summers of 2007 and 2012 have been widely reported, as well as the modest increases in winter ice extent in the southern hemisphere. Volume changes for Arctic sea ice are more consistent than changes in extent (Lindsay et al. 2009), and volume changes are a more reliable and earlier diagnostic of sea ice change than ice extent in global climate models forced by increased greenhouse gases (Schweiger et al. 2011). In the Antarctic, there is strong regional variability in sea ice trends but an overall increase in winter sea ice cover (Stammerjohn et al. 2008), both thought to be a response to the changing wind field and also due to ENSO teleconnections. Changes in stratification of the surface mixed layer due to increased precipitation or glacial melt may also play a role in reducing bottom melt rates and hence create a mechanism for increasing the winter ice cover (Zhang 2007). The differing evolution of ice extent in the two hemispheres is undoubtedly related to the different geographical configurations at the two poles. However, determining the specific mechanisms driving the trends in each hemisphere is still an open research question.

Spectacular disintegrations of several ice shelves have occurred over the past few decades along the Antarctic Peninsula. As a result, the outlet glacier feeding these ice shelves has sped up leading to accelerated ice loss (Scambos et al. 2004). In Greenland, a complicated pattern is observed in the outlet glacier behavior within the past decade. While a number of glaciers have experienced retreat, thinning, and acceleration, some have experienced readvance, thickening, and deceleration (Moon et al. 2012; McFadden et al. 2011). Polar ice sheets have contributed significantly to global sea level rise (Shepard et al. 2012). Increased sub-ice-shelf melt rates caused by interactions with warming oceans will result in a strengthening of the ocean stratification that could reduce vertical mixing of heat and gas and inhibit convection from occurring in the polar seas. This impact would have far-reaching climatic implications in the large-scale overturning circulation of the North Atlantic and global oceans. At present there is very little understanding of this process as there are too few observations.

Future research

Logistical difficulties and the harsh environment have meant that polar regions have historically been poorly observed and therefore many fundamental physical processes are still not well understood. Changes in the poles are occurring rapidly and it is imperative that we have an adequate sustainable monitoring network in place to detect and quantify these swift changes. Given the sparse amount of polar measurements that have been collected using a variety of instruments by different nations at different times and in different regions, there is also a real need to synchronize the quality control and processing of these data. The coordinated treatment of the existing datasets will allow for consistency and ultimately produce better uncertainty and bias estimates. Observations of air-sea heat, freshwater, and gas fluxes in polar regions represent enormous instrumental challenges to withstand harsh conditions and associated maintenance constraints. Inaccuracy in these fluxes has a significant impact on both observational and modeling studies. Finally, the continued operation of satellite and airborne observing missions is of vital importance to monitor ongoing and future changes of the ice sheets. Complemented with field campaigns, these observations are indispensable to improve our understanding of the physical processes and mechanisms that control and drive the ice sheet changes.

Despite the crucial role of the Southern Ocean in the Earth system, understanding of particular mechanisms involved remains inadequate. In particular, it is becoming increasingly likely that mesoscale eddies play a critical role in maintaining the Southern Ocean stratification (Marshall and Radko 2003) and governing its response to changing winds (Hallberg and Gnanadesikan 2006), as well as in the meridional transport of heat, salt, and biogeochemical tracers, such as carbon. These effects are complex and remain poorly understood. In particular, the eddies are not routinely resolved in climate models and are instead parameterized. These parameterizations are often derived empirically that in the past have been largely unverified through process studies, although recent targeted efforts such as the US CLIVAR DIMES program will provide some future guidance. Progress in understanding the effects and dynamics of eddies is urgently needed so as reduce uncertainty in the predictions and improve our understanding of the importance of the Southern Ocean in a changing climate.

While observations are required to validate model predictions and improve our understanding of the basic mechanisms, in turn modeling efforts can also be used to simulate processes that are difficult to observe and thus are poorly understood. A coordinated effort is particularly important in polar regions where remote locations make it difficult to undertake observations and where different models currently provide different realizations of polar characteristics and processes. A concerted integrated research effort is needed to identify what processes the models need to capture to improve simulation and predictability of change in the polar regions.

5.4 Climate and marine carbon/biogeochemistry

Climate is not just a matter of physics. Biogeochemical processes impact the chemical composition of the atmosphere and oceans, as well as exchanges between them. The increase of atmospheric carbon dioxide (CO_2) and global warming are subject areas currently of great societal importance. The global ocean, as a major sink of anthropogenic CO_2, significantly slows CO_2 increase in the atmosphere. At the same time, increased greenhouse warming from CO_2 affects the ocean's biogeochemistry and ecosystems in complex and uncertain ways. The ocean similarly impacts distributions of other important elements (e.g., nitrogen and sulfur, the latter potentially a source of nuclei for cloud condensation). Developing a better understanding of the interplay between climate and the chemical and biological properties of the world's oceans is a formidable problem, but the stakes in not doing so are enormous.

Definition

Interactions between climate and biogeochemical processes encompass a broad range of processes in terms of type and scale. It is impractical for a single program to cover the topic comprehensively. The niche that US CLIVAR can fill relates to one of its core strengths: determining how variations in climate mediate air-sea fluxes, and ultimately the chemical states of the atmosphere and ocean. Details are included below on three specific topics within this overall Research Challenge: marine ecosystems, carbon cycling, and the Southern Ocean. US CLIVAR is poised to be both a resource for, and an active participant in, the research being done in these and other related areas.

Significance

Marine ecosystem

The marine ecosystem is societally important in many ways, providing a critical food source for the rising world population, as well as cultural and recreational benefits. Climate variability is known to have a large impact on these goods and services. International management activities are now in place, or soon will be, to oversee the health of the marine ecosystem in most of the world's oceans. A better understanding of the present and future states of marine ecosystems on climatic timescales is essential to these management efforts, and will certainly lead to substantial societal payoffs (e.g., help to limit overexploitation).

Carbon cycling

The global ocean has historically been a major sink of anthropogenic CO_2. However, global warming is changing the ocean circulation and its ability to uptake CO_2 by altering winds and surface heating, and these changes are expected to accelerate in the future. Changes in oceanic carbon cycling and biogeochemistry may feedback positively onto the atmospheric carbon dioxide concentrations through the slowdown of oceanic carbon uptake, further enhancing global warming (Friedlingstein et al. 2006). The strength of these feedbacks depends on the complex interplay between physical and biogeochemical processes regulating the sensitivity of ocean carbon uptake to climate perturbation. Uncertainty in the fate of the oceanic carbon sink is a major source of uncertainty in understanding future climate, due to the number of processes and the time and spatial scales involved (Orr et al. 2001).

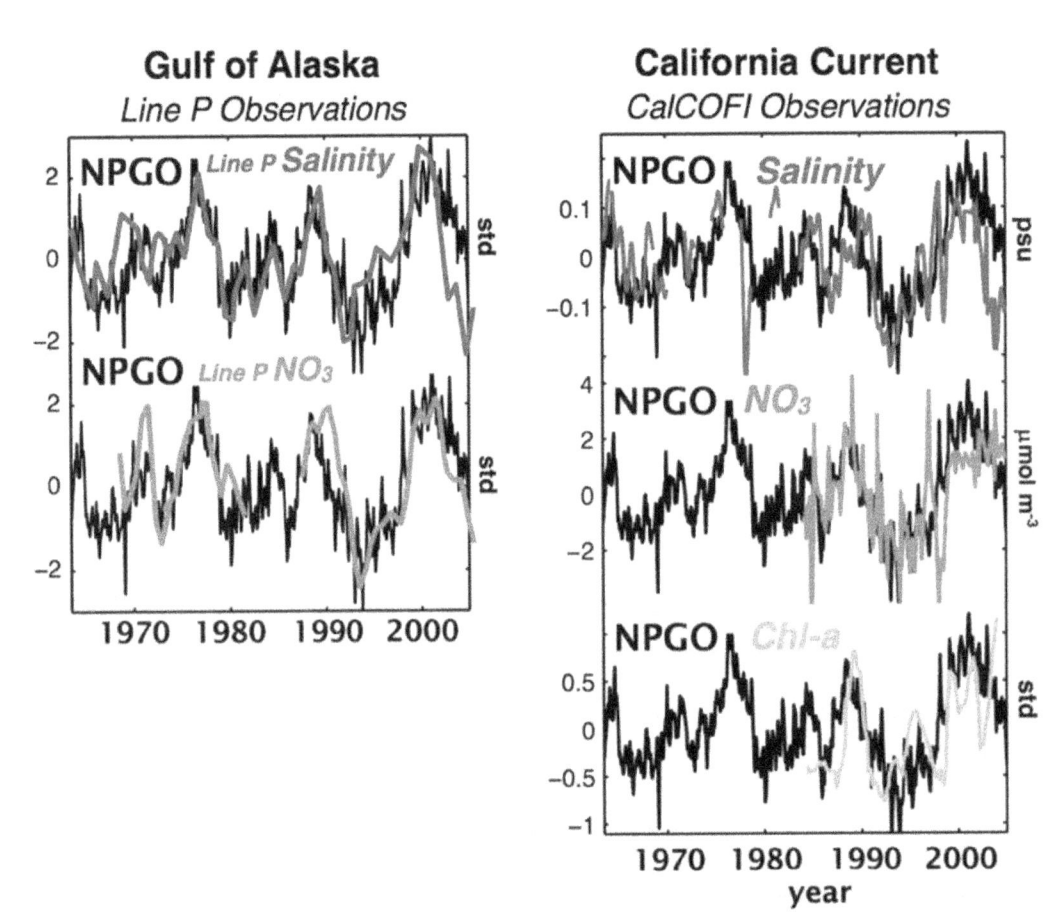

Figure 5.3

Time series of NPGO index (black) is compared with anomalies in salinity (blue), nitrate concentration (purple), and chlorophyll-a (green) recorded in long-term observations in the Gulf of Alaska and California Current. All times series are plotted in standard deviation units (std) except for nitrate concentration (NO_3 in mole/m³) and salinity (psu). The chlorophyll-a time series is smoothed with a 2-year running average. The close connection between the NPGO index and biological variables is apparent. See Di Lorenzo et al. (2008, 2009) for details.

Dynamics

Marine ecosystem

The response of marine ecosystems to climate variability and change is difficult to predict, given the multiple, often confounding factors that affect it. Phytoplankton growth reacts strongly to changes in the physical environment in many locations, particularly to the thickness of the surface mixed layer and to the processes that lift subsurface nutrients into the euphotic zone. An example is shown in Fig. 5.3, with the North Pacific Gyre Oscillation representing an index for horizontal advection and the upwelling by the regional winds. It is unclear, however, whether these kinds of relationships are robust to a changing climate. Marine ecosystems are sensitive to changes in upper ocean stratification and ocean chemistry. There is particular concern that some lower-trophic level species that are an important source of prey (e.g., pteropods) may be severely disadvantaged by ocean acidification. The complexity of most marine ecosystems further implies that they will respond nonlinearly to climate variability, with reorganizations in community structure and novel sensitivities (or the lack thereof) to drivers that were previously important. Finally, climate change impacts may include deep-sea communities (Smith et al. 2009) and have the potential to significantly alter the biological carbon uptake.

Carbon cycling

Likewise, the climatic response of marine carbon cycling is also difficult to predict. Existing coupled carbon-climate models suggest that the inclusion of an interactive carbon cycle increases the sensitivity of climate to a given emission rate. This happens because both terrestrial and oceanic carbon sinks become less effective over time in a warming climate, but there is considerable uncertainty about the strength of these feedbacks in both models and observations. Further, climate models have shown that a substantial fraction of model-to-model differences in biogeochemical fields results from the propagation of known errors in model physics (Doney et al. 2004; Matsumoto et al. 2004); as a result, there is still considerable uncertainty about the physical mechanisms responsible for carbon uptake, and their relative importance in different regions and at different timescales. For example, an increase in ocean stratification can alter the rate of vertical exchange of dissolved inorganic carbon and nutrients in the water column, whereas an increase in wind forcing may lead to more mixing and upwelling thereby achieving the opposite effect. A slowdown of the AMOC can significantly reduce carbon uptake in the northern North Atlantic (Sarmiento et al. 1998), whereas weaker upwell-ing in the equatorial Pacific will reduce the outgassing of CO_2 in that region (Lovenduski and Ito 2009)

Southern Ocean

As a further complication, the Southern Ocean (south of 30oS), the least measured and possibly understood basin of all, plays a crucial role in the ocean carbon cycling. Alone, it may account for up to half of the annual oceanic uptake of anthropogenic carbon dioxide from the atmosphere (Gruber et al. 2009); moreover, the vertical exchanges that take place in the Southern Ocean are responsible for supplying nutrients that fertilize three-quarters of the biological production in the rest of the global ocean (Sarmiento et al. 2004). A wind-stress increase over the Southern Ocean can increase the rate of regional deep-water upwelling (Toggweiler and Samuels, 1998), and this increase will enhance both the outgassing of natural CO_2 and the uptake of anthropogenic CO_2 (Lovenduski et al. 2008). This is because deep waters in the Southern Ocean are enriched in natural CO_2, and so upwelling will intensify outgassing; at the same time, because deep waters are also colder, they are more prone to the uptake of anthropogenic CO_2 when they reach the ocean surface. In the current climate, the net CO_2 exchange is dominated by the natural component, but the balance under global warming is unclear.

Future research

Although the processes that maintain the marine ecosystem and account for CO_2 exchange between the ocean and atmosphere are qualitatively understood, existing coupled physical/biogeochemical models are not yet able to simulate them accurately. Thus, understanding how they will respond to climate variability and change is problematic. A key challenge then is to increase our understanding of the coupled physical/biogeochemical processes that maintain the marine ecosystem and oceanic sources and sinks of carbon and accurately predict how they will evolve in response to climate variability and change.

A multi-faceted approach is required to address these issues. It is summarized in the following four activities:

Multi-purpose and integrated ocean-observing networks:

There should be a continuing emphasis on the development and deployment of multi-purpose ocean-observing systems, as endorsed by the Ocean Obs'09 Framework for Ocean Observing (Lindstrom et al. 2012). Such a course will not only enhance the efficiency of the ocean-observing network, i.e., the bang for the buck, but will serve to foster collaboration between scientists of different sub-disciplines. The establishment of these observing systems may include not just technical challenges but also complications due to financial support being required from multiple sources. In this regard, US CLIVAR could play a key planning role. An example of a present monitoring effort that could be enhanced is the CLIVAR/CO_2 Repeat Hydrography Program. The inclusion of more biogeochemical measurements with this program would help in the diagnosis of the various processes that determine the fluxes at the air-sea interface and ultimately the carbon inventories impacting the climate.

Continued innovation of oceanographic instrumentation: In recognition of the substantial costs in collecting oceanographic observations from research ships, there should be effort devoted to developing ways to take advantage of ships of opportunity and the use of autonomous platforms. In the former category, there are now long data records based on continuous plankton recorders (CPRs) and increasing use of automated sampling systems. Outfitting these systems with the capability to sample chemical and biological properties is becoming more technologically feasible. There is a trend towards the use of more autonomous platforms, in particular sub-surface and wave-powered gliders. Greater use of these instruments, especially when equipped with enhanced measurement capabilities, will allow better specification of the ocean on the temporal and spatial scales that are important to the ecosystem.

Integrated ecosystem process studies: These are generally major and expensive undertakings, but will continue to be valuable and in some cases essential for gaining mechanistic understanding of the biological responses to climate variations and change. From an agency perspective, it may be difficult to initiate and manage multi-disciplinary programs, but there are previous examples (e.g., the GLOBEC program) for which this was accomplished successfully, and from which lessons learned can be drawn upon.

Coupled physical/biogeochemical modeling: Interactions between the climate, the carbon cycle, and marine ecosystems can be modeled through statistical/empirical or dynamical approaches. The two types of models each have their benefits and drawbacks for collaborative modeling approaches as discussed in Chapter 4. The challenges in modeling ecosystem interactions suggest the utility of multiple approaches. US CLIVAR should encourage development, evaluation, and comparison of different type (statistical, dynamical, and hybrid) models. Some reason-

ably long and comprehensive datasets exist, for example from the CalCOFI program and for Georges Bank, which may be sufficient for this purpose. CLIVAR may also be in a position to support the improvement of model interoperability. Conceivably the difficulties involved in coupling physical to biogeochemical and socioeconomic models are limiting the amount of vertically-integrated modeling that is being done.

Chapter 6
Cross-Cutting Strategies

Specific types of activities are needed to achieve each of the US CLIVAR science goals (Chapter 4). They can be grouped into five distinct elements (Cross-cutting Strategies): (1) sustained and new observations; (2) process studies; (3) model development strategies; (4) quantifying improvements in predictions and projections; and (5) communicating climate variability. Table 6.1 provides a summary of how the cross-cutting strategies outlined in this chapter address each of the US CLIVAR Goals. This chapter describes the Strategies, and provides specific examples of research activities within each of them that support US CLIVAR goals.

6.1 Sustained and new observations

Observing and understanding climate variability and change requires sustained monitoring systems spanning decades. Documenting ocean variability, particularly on decadal timescales, is critical for understanding how the ocean regulates the climate system and how it responds to climate change and other anthropogenic influences. Observing how the ocean interacts with the atmosphere, the cryosphere, and terrestrial systems is needed to understand changes in Earth's ecosystems and habitability. Long-term observational records provide data to evaluate climate models and allow us to better quantify uncertainties in their predictions and projections. New observational capabilities are required to fill gaps in the existing climate observing system and to apply new technologies to the sampling of aspects of the climate system that have previously not been accessible.

Sustained observations

An effectively designed and sustained climate observational network can only be achieved with long-term commitments.

Sustained climate observations, with careful calibration and validation over time, are crucial to understanding and observing climate variability and change (Houghton et al. 2012) and to test hypotheses about key controls on climate. Keeling's sustained observations of atmospheric carbon dioxide, originally designed to quantify the capacity of the atmosphere to carry CO_2, ultimately has led to perhaps the single most important climate record. The record documenting the persistent decline in Arctic sea ice has been critical to understanding long term climate change in the Arctic, but began as a byproduct of NASA's Nimbus-7 satellite missions. The role that the Tropical Ocean-Global Atmosphere (TOGA) project played in building the TAO array in the Tropical Pacific was critical for providing long-term ocean and lower-atmosphere observations of ENSO (McPhaden et al. 2010). These three examples of long-term climate records originally began as research measurements that were transitioned into operational commitments. US CLIVAR will help agencies identify and plan for such transitions.

Consistent with Global Climate Observing System (GCOS) goals and the OceanObs'09 call to action (Fischer et al. 2010, Lindstrom et al. 2012), US CLIVAR calls for continued support to sustain ongoing collection of key or essential climate variables at key locations. Capabilities must be sustained through the next decades to measure Earth's radiation budget and atmosphere-ocean interaction on the timescales of climate variability and anthropogenic climate change. Satellite ocean surface wind vectors and SST measurements provide the key upper boundary conditions on the ocean circulation. Ocean station time series from buoys and moored arrays (e.g., Ocean Reference Stations, TAO, PIRATA, and RAMA arrays) provide continuous measurement and monitoring of the upper ocean, lower atmosphere, and air-sea interaction at the interface of climate variability on daily to decadal timescales. These observations are important for understanding and predicting ENSO and decadal modes of climate variability such as the PDO and AMO.

Upper ocean property profiles from the broadscale Argo array and repeat high-density XBT transects are needed to measure the mass and heat storage and transport that drives decadal ocean circulation variability. The US Global Ocean Carbon and Repeat Hydrography Program that provides measurements of carbon and other biogeochemical observations needs to be maintained and strengthened. *US CLIVAR activities advocate and leverage these long term climate monitoring strategies to better understand and predict climate variability. US CLIVAR also encourages extension of recently acquired capabilities to measure processes of climate variability on seasonal to interannual timescales, to the decadal and multi-decadal scales of climate change.*

New observations

There are many aspects of the climate system that are currently not well observed and there are gaps in the existing networks that require new observational capabilities. New observational capabilities are needed to understand biogeochemical changes, to better quantify the sources and sinks of carbon, and to constrain aerosols, clouds, and climate feedbacks. *Increasingly, new observational opportunities cross traditional disciplinary boundaries, and so US CLIVAR will work with other scientific communities to build the next generation of climate observations.* This requires technological improvements and improvements in sampling strategies. The development of profiling floats that led to the global Argo array is an example where improved technology combined with support for widespread deployment catalyzed a revolution in upper ocean sampling.

Cross-Cutting Strategies⇒ Goals ⇓	Sustained and new observations	Process studies	Model development strategies	Quantifying improvement in predictions and projections	Communication of climate information
Understand the role of the oceans in observed climate variability on different timescales	Document var at ons	Data to eva uate and mprove mode s	Improve mode ng of c mate across processes and t mesca es	Understand m ts of c mate pred ctab ty	Pr or t ze observ ng network and pred ctab ty stud es and mprove pred ct ons of ocean and c mate var ab ty
Understand the processes that contribute to climate variability and change in the past, present, and future	Document c mate-cr t ca processes	Invest gate processes to he p exp a n var at ons	Property conserv ng c mate reana yses	Quant fy ng mportance of mode uncerta nty n project ons	Set pr or t es for observat ons and pred ctab ty stud es; commun cate about conf dence and pred ctab ty
Better quantify uncertainties in the observations, simulations, predictions, and projections of climate	In t a ze and eva uate mode s mu at ons	Mode assessment	Improve mode s	Quant fy mode , ntr ns c and scenar o errors	Address needs for pred ctab ty and sens t v ty stud es
Improve the development and evaluation of climate simulations and predictions	In t a ze and eva uate c mate mode s	Prov de data to deve op and test mode process representat on	Reduce b ases n c mate mode s	Quant fy mportance of mode phys cs errors	Determ ne key targets for mode deve opment across commun t es
Collaborate with research and operational communities that develop and use climate information	Prov de mu t-d sc p nary datasets	Prov de process understand ng and opportun ty for co aborat on across d sc p nes	Commun cat on between observat ona and mode commun t es	Improved commun cat on across d sc p nary boundar es	Prov de nformat on on dom nant c mate phenomena and pred ctab ty

Table 6.1 *Cross-cutting strategies and their intersection with US CLIVAR goals.*

US CLIVAR will seek to identify and call attention to climatically important but currently undersampled regions. Two such examples are the deep ocean, which is critical for understanding decadal climate variability and the distribution of energy in the climate system, and high latitude regions, where observing climate system change and its contributing factors has been particularly challenging. Each region is briefly discussed next along with possible ways in which US CLIVAR will contribute.

The deep ocean

During OceanObs'09, Rintoul et al. (2010) identified the deep ocean as one of the undersampled ocean regions that is key to climate variability and observations are needed to address a wide range of critical climate questions. Specifically, Rintoul et al. (2010) called for the maintenance and enhancement of established sites and technologies, including moored arrays, full-depth hydrography, Argo sampling of the upper ocean, and satellite observations (altimeter and gravity) of depth-integrated quantities. They also called for the study and development of new observing technologies, including deep-profiling Argo floats, long-duration gliders capable of sampling deep-ocean properties, as well as the use of acoustic tomography and thermometry to provide integral constraints on ocean heat content and ocean currents. *US CLIVAR will contribute to the development of new and sustained deep-ocean observations by highlighting where deep-ocean data gaps exist. This can be achieved through a specific US CLIVAR focus on deep ocean observations in conjunction with the US Global Ocean and Carbon and Repeat Hydrography Program.*

High latitudes

Polar oceans are another climatically important sector of the ocean that remains grossly undersampled. Due to the historic paucity of *in situ* observations in the high-latitude seas, only limited attention has been given to hypotheses involving ocean variability that may originate in polar regions, e.g., changes to ocean vertical stratification, advective heat transport, and heat content. Although the number and quality of high-latitude *in situ* ocean observations has dramatically increased during the past ten years, little progress has been made towards quantifying the role of ocean variability in the observed changes of sea ice thickness and extent. The polar seas are still relatively sparsely sampled in space and time with respect to the processes important for sea ice-ocean interaction. US CLIVAR will continue to facilitate the investment of resources in high latitudes. US CLIVAR will benefit from interaction with programs such as the

NSF Arctic Observing Network and the NSF Ocean Observatories Initiative (OOI) that plan long-term (~25 year) moored arrays combined with gliders in the Irminger Sea, in the Argentine Basin, in the South Pacific, and enhancement of the array at NOAA site Ocean Weather Station Papa. There is also a great need for expanding and creating observational networks that address key ocean-related climate issues in the Southern Ocean (see section 5.3). Expansion of the profiling float arrays under sea ice and the aggressive use of gliders in regions where the Antarctic ice sheet is especially vulnerable to incursions of warm water are needed to understand sea ice-ocean interactions in the Southern Ocean. US CLIVAR advocates a specific focus on polar ocean observational data gaps and how they can be mitigated with the development, deployment, and coordination of additional high latitude ocean observing systems.

6.2 Process studies

Focused studies of key processes underpinning the climate system are critical for understanding the ocean's role in climate. Process studies provide quantitative understanding of the mechanisms controlling climate change and variability and provide observational data to evaluate and improve models. Model uncertainty is the largest source of error hampering accurate prediction of decadal climate variability, and internal variability of the climate system is the primary impediment to accurate prediction of intraseasonal to interannual climate (Hawkins and Sutton 2009). By targeting aspects of the climate system that are poorly understood and represented in models, process studies provide the data to better quantify model uncertainties and to improve models.

As described in the Accomplishments Report, US CLIVAR has played a central role in process study design and implementation by establishing a set of best practice guidelines (Cronin et al. 2009) aimed at fostering improved communication between observational and modeling communities and improving data archiving and dissemination. US CLIVAR has identified key gaps in process understanding that are hindering our ability to predict climate variability and change. *US CLIVAR will continue to ensure that data and process-level understanding gained from process studies is optimally used to benefit climate model evaluation and development.*

The atmosphere and ocean are turbulent systems with poorly-understood dynamics encompassing a broad range of time and space scales. Basic processes such as ocean mixing, cloud and precipitation formation and air-sea, land-air and air-sea-ice physical, chemical, and biological interactions occur on smaller spatial scales than will be resolved by most climate models over the coming decades. It is well-recognized that these basic processes and their co-interactions must be represented in climate models to achieve accurate representation and predictions of the climate system. For example, atmospheric convection in the tropics plays a fundamental role in determining the strength and propagation of the Madden-Julian Oscillation (Zhang 2005); without accurate representation of subtropical low clouds, models are unable to accurately represent the spatial distribution and seasonal cycle of tropical precipitation (Yu and Mechoso 1999); accurate representation of ocean mixing in climate models is critical for reproducing the major overturning circulations in models (Wunsch and Ferrari 2004); the representation of sea ice is critical for understanding Arctic climate variations (Johannessen et al. 2004). Intensive field studies provide much needed observational data to understand these basic physical, chemical, and biological processes at the space and timescales on which they occur.

In addition to intensive studies of basic processes, a relatively new category of observational climate process study is emerging that encompasses larger temporal and spatial scales than most field studies but which requires enhanced regional observational infrastructure for periods of months to several years. As discussed below, these studies allow a deeper exploration of key mechanisms than can be provided by the standard observing network. As reanalyses and the models and global observing system used to generate them improve, process studies will increasingly benefit from these datasets. Indeed, there is an important opportunity to increase the coordinated use of such auxiliary information in the design of process studies not just in their analysis. In addition, process studies are increasingly focusing upon the connections between physical, chemical, and biological systems. *US CLIVAR will play an important role in coordination and collaboration across disciplines for process study design.*

Intensive field studies

New intensive field studies are needed to provide basic process level understanding and physical parameterizations of turbulent processes in the ocean and atmosphere. The small-scale turbulent nature of the ocean and atmosphere, not sampled by the global observing network, renders parameterization of mixing processes in both fluids critical for the accurate representation of large-scale flows, clouds, gas and particle transports, and biogeochemical processes in climate models. Intensive field studies are helping to fill gaps in process understanding but more are needed. *US CLIVAR will ensure that the climate model development community is closely involved in the early stages of process study design to ensure that the data to be collected will maximally benefit model improvement and parameterization development.*

Interactions between the physical, chemical, and biological processes are of critical importance to current and future climate. The amount of sunlight reflected by marine low clouds is strongly dependent upon the quantity and size of marine aerosols, many of which are now thought to be derived from biological and biochemical processes in the ocean (Mahowald et al. 2011). Marine aerosols determine the sensitivity of the Earth's radiation budget to perturbations in aerosols associated with anthropogenic pollutants (Hoose et al. 2009). Ocean biogeochemistry is central to ocean carbon dioxide uptake (Sabine et al. 2004). Future process studies will need to focus on these interactions. *US CLIVAR will advance such studies by fostering new communications with the climate-relevant parts of the chemical and biological communities. Interagency and international coordination provided through US CLIVAR will optimize the use of observational platforms to concentrate efforts.*

Climate phenomena studies requiring augmented monitoring

Observational studies are needed to determine the sources and limits of predictability of climate phenomena such as regional monsoons and the MJO and to understand key aspects of regional ocean circulations and their interaction with ice sheets. These studies are conducted on time and space scales longer than those of typical intensive field studies, but nevertheless require significant investments in temporary observational infrastructure. Interagency coordination is essential to delivering sufficient observational infrastructure to meet the needs of each particular study. Examples of climate process studies include the North American Monsoon Experiment (NAME, Higgins et al. 2006), the Inter-Americas Study of Climate Processes (IASCLiP), the Diapycnal and Isopycnal Mixing Experiment in the Southern Ocean (DIMES, Meredith et al. 2011), and Year of Tropical Convection (YoTC, Waliser et al. 2012). Such studies often incorporate intensive field phases but also aim to enhance the observational network or generate focused specific climate datasets (e.g., using satellites or enhanced meteorological analyses) for a period of months to years and at the continental/basin scale. Key locations in need of enhanced climate process studies are the Arctic, where additional measurements are needed, for example, to understand the role of clouds in the surface energy balance, and the Southern Ocean, where interactions between the atmosphere and ocean are currently poorly monitored. The new NSF Ocean Observatories Initiative and the DoE Atmospheric Climate Research Mobile Facilities are examples of new facilities that will provide high quality process level data for periods of months to years. *US CLIVAR will ensure that the community takes full advantage of these new facilities by integrating them into the design of climate phenomena studies.*

Future process-study design

A great opportunity exists to improve the design of intensive process and climate phenomena studies by making better use of model simulations to optimize the deployment of fixed and mobile observational platforms. Field and climate process studies have traditionally relied upon human judgment to determine the optimum configuration of observations deployed in process studies. In the last decade the satellite community has been transformed by the increasing use of Observing System Simulation Experiments (OSSEs) to allow quantitative assessment of the value of an observing system for addressing particular needs (Atlas 1997). OSSEs can also be used for estimating the limits to which an observing system can determine the current oceanic state and, therefore, for assisting users of data in interpretation of data and development of analysis techniques. In designing an ocean mixing tracer-release experiment, an ensemble of regional ocean models could be run as an OSSE to predict tracer evolution, and then ship, mooring, and float sampling could be optimized accordingly. A limited application of such an approach was attempted prior to the tracer release and float deployments as part of the U.S CLIVAR DIMES experiment (see Accomplishments Report). The ever-expanding computational power available will allow such simulations to encompass the range of scales important to field and climate process studies. In turn, the process observations will help evaluate and improve the models themselves and aid in the assessment of internal variability and its predictability. *US CLIVAR will develop strategies for using ensembles of eddy-resolving ocean simulations and cloud-resolving regional simulations to optimize the use of observational platforms to collect data in process studies and to maximize the use of and coordination with the global land-based and satellite observing system.*

6.3 Model-development strategies

Climate models and other models of the Earth system or its components are critical tools for modern climate science. These models are used for a variety of purposes, the most widely known being applications to numerical weather forecasting and future climate projection. The choice of model depends upon the type of application. Different models are used for prediction, forecasting, projection, scenario development, and experiments (see Chapter 4). Model development efforts in each of these different applications vary substantially, but numerical climate modeling underpins all applications. Models are also central to the development of better reanalysis datasets, which help us to understand climate variability and controlling processes. Climate model representation of physical, chemical, and biogeochemical processes in the Earth system has improved dramatically since US CLIVAR's inception, but many processes are still poorly represented, and this limits their skill. *US CLIVAR will address its model development and evaluation goal by fostering better communication and practices between model development and observational communities.*

US CLIVAR encourages using novel model strategies such as coupled modeling and modeling with hierarchies of models of differing complexity. Coupling the ocean and atmosphere is the essential component of a climate model. Now biogeochemical, ecological, and economic models are being integrated into climate modeling. Coupling each new component to the climate system increases the complexity of the problem. Hierarchies of models of differing complexity help us understand behavior of complex or chaotic systems. These sorts of techniques were developed effectively under US CLIVAR to understand ENSO variability.

Climate Process Teams

Climate Process Teams (CPT), developed within US CLIVAR (see Accomplishments Report), were designed to speed development of climate models by bringing together model development specialists with observationalists and process modelers to focus on the most critical model deficiencies. The benefits of CPTs span all five of the US CLIVAR goals. CPTs have a strong model development component, but by fostering the synthesis of new observational datasets from process studies and satellites, process level modeling, and model assessment, they also provide an avenue for new scientific discovery at the interface between modeling and observation.

Process observations themselves are necessary but insufficient requirements for model development because such observations can generally constrain only a few weeks of predictions on a local scale. To use process studies to improve a model permanently, the observations must be understood as sampling the behavior of a climate process. Then the understanding of that process must be simplified and represented in a mathematical form (a parameterization) that can be numerically implemented as a part of a climate model. As noted, new or improved parameterizations are often a key result of a CPT. The development of parameterizations is best served by a team of observationalists, theoreticians, and modelers working together closely. *US CLIVAR, having played a key role in initiating the CPTs, provides critical oversight on the progress of the current teams.*

Because of their success (see Accomplishments Report), the CPT concept should be continued and all agencies with a mandate to foster climate science should be encouraged to collaborate in funding future CPTs. According to the best practices that have been identified for successful CPTs, future CPTs should focus on processes that are currently poorly represented in climate models where there is a clear pathway for improvement in representation that may lead to better climate simulations. Successful CPTs should be highly focused on a single process or set of closely interlocking processes and should have application beyond a single model. Future CPTs will take advantage of opportunities driven by the combination of new process studies, global observational datasets, and computational resources. The current CPTs are funded by NSF and NOAA, but in the future other agencies should be encouraged to fund CPT projects. Candidate processes in need of CPTs include tropical atmospheric convection (Arakawa 2004), deep overturning ocean circulations, ice-ocean dynamics (Kirtman et al. 2012), and biogeochemical processes involved in ocean carbon uptake (Riebesell et al. 2007). Agencies beyond NSF and NOAA have significant interests in these areas, and *US CLIVAR will help facilitate interagency collaboration to support future CPTs.*

Data assimilation and seamless prediction
Numerical weather prediction uses data assimilation to incorporate a wide variety of available observational data to improve the accuracy of forecasts. Models used for climate projections, on the other hand, tend not to include detailed data or observations during the course of the run, except as initial conditions or as a basis for model evaluation. There is great potential, however, in having projection models capable of true data assimila-

tion and numerical weather models capable of simulation over longer timescales. "Seamless prediction" (Palmer et al. 2008) is the process of developing models capable of being used at both short and long timescales without drastic differences in model code. It is presently unknown whether seamless prediction modeling efforts produce better forecasts or projections than model development focused on either the short or long timescales. However, it is clear that seamless prediction offers real potential for model development by incorporating techniques from both weather and climate communities, thus providing the impetus for improved communication between the different communities and for technology sharing.

The US CLIVAR goals (Chapter 4 and Table 6.1) emphasize the development and communication between these different communities of modelers and observationalists. Conventions regarding data storage, model development, and evaluation protocols need to be shared between these communities. *US CLIVAR will play an increasing role in strengthening connections between the climate and weather model evaluation and development communities.* On short timescales, data assimilation and predictions are ideal for understanding the role of the oceans and processes (Goals 1-2). Ocean data assimilation is becoming an increasingly important aspect of seasonal to interannual coupled climate forecasting, and US CLIVAR is in an ideal position to ensure that developments in assimilation are widely distributed and adopted by other communities that would benefit from the information. Even if seamless prediction never proves as accurate as dedicated forecast models, the learning and collaboration between the forecasting and climate modeling communities needed to develop these models will help achieve US CLIVAR goals.

Virtual Earths and idealized modeling
Virtual Earth modeling involves simulating the Earth system as best as possible and then perturbing the system (e.g., by closing the Drake Passage or by removing mountain ranges) to learn about how key aspects of the Earth's climate system function. Idealized modeling and the use of "virtual Earths" for experimental numerical modeling is less common in US CLIVAR projects than is more realistic modeling. However, idealized techniques are often a valuable part of establishing process-level understanding by simplifying complex phenomena. There is a need for a hierarchy of models that span a broad range of complexity (Held 2005). Idealized models are always subject to the error of incorrectly simulating true scenarios. *However,*

US CLIVAR will aid in the cross-communication of the design of idealized simulations so that observations and the scientists who collect them have an opportunity to contribute to the discussion about whether idealized models are being used in the correct regime. Furthermore, the results of these simulations can be a useful tool for building intuition and the ability to recognize the signatures of certain processes in more complex data.

Experimenting with idealized models can greatly improve understanding of the role of the oceans, processes, and predictability across timescales (Goals 1-3). Such experimentation often helps to establish the sensitivity of models, which in turn helps us understand what steps are effective in developing future models (Goal 4). Finally, these tools are ideal for collaborations involving the broader physics and mathematics communities, who have much to offer but are often not well-connected to the climate problem (Goal 5).

Empirical and statistical modeling

Sometimes processes are too poorly understood, too complex, or occur on scales too small to directly handle computationally. If, however, a numerical model is still required for forecasts or projections, then sometimes an empirical or statistical model can be developed instead of a model based on physical principles. An empirical model is one where scaling relationships between observed variables are used to project future behavior. Typically, these models are not yet proven to be connected to any underlying physical principles. A statistical model is one where observed scaling relationships and physical principles are incorporated simultaneously but in a non-deterministic way, so that the predictions of the model incorporate some estimate of the uncertainty stemming from the quality of the observations, the fidelity of the model, and the intrinsic uncertainty of modeling a chaotic system.

Both empirical and statistical models succeed largely on the quality of data that they are designed to reproduce and secondly on the statistical modeling techniques and methods of data assimilation and parameter estimation used to construct them. Often these models outperform more complete Earth system models in forecasts or projections of a limited nature, especially when asked to predict those metrics that were used to create them. Like other models described above, statistical models benefit from consistent data format, availability, quality control, and techniques of model evaluation.

These empirical models help to address a number of US CLIVAR goals. They can act as competition or bases for the evaluation of physical models (Goal 4) and can assist in identifying key processes. They can help to quantify uncertainty and predictability, even if the underlying processes are not understood (Goal 3), or be used in experimentation to better understand the role of the oceans across timescales (Goal 1). Finally, they can be shared in collaborations with research and forecast communities to build understanding of key processes and their contribution to variability and change (Goal 5).

6.4. Quantifying improvements in predictions and projections

The current generation of climate models represents a significant advancement in sophistication over the previous generation. Improvements include increases in spatial and temporal resolution, and better representation of physical processes including ocean dynamics, air-sea coupling, land-surface modeling, land-atmosphere coupling, stratospheric processes, and snow and sea ice modeling. These improvements should lead to improvements in the quality of predictions and projections, but it is important that techniques are developed to critically assess these improvements in order to build the confidence of users and to identify the most likely targets for future improvements.

State-of-climate predictions

Over the last few decades there has been significant progress in our ability to predict climate variability on seasonal and intraseasonal timescales by capitalizing on ocean-atmosphere coupling associated with ENSO and MJO, but understanding the limits of intraseasonal and seasonal prediction skill is needed. Understanding forecast quality and the limitations of forecast accuracy are important for addressing a number of US CLIVAR goals and for understanding decadal predictability.

Besides short-term climate variability, ocean-atmosphere coupling is a primary driver of decadal climate variability (see section 5.1). The presence of decadal modes of climate variability does not guarantee predictability of the modes at those timescales, especially as there is evidence that some part of the decadal variability is driven by changes in anthropogenic aerosols (Kaufmann et al. 2011, Booth et al. 2012), stratospheric water changes (Solomon et al. 2011), and other factors not directly linked to ocean-atmosphere coupling. *However, based on idealized studies it is likely that predictability exists, and US CLIVAR encourages activities to identify and exploit it.* Because currently-available observational records are short, especially for the ocean, we must develop a variety of strategies to study and quantify decadal predictability. Such tools include coupled climate models, the fidelity of which is continuously evaluated (Collins et al. 2006; Furtado et al. 2011), and linear inverse models (e.g., Newman 2007, Pegion and Sardeshmukh 2011). Exploitation of significant rapid climate shifts between decadal phases is showing some considerable promise for quantifying and evaluating coupled model predictions (e.g., Robson et al. 2012).

Increasing greenhouse gases (GHGs) also provides predictability on decadal timescales. Extrapolation from recent temperature trends or GCMs forced with increasing GHGs can provide predictions on decadal scales, but natural decadal variability is expected to compete with secular climate change trends for the next several decades, especially at the regional scale (Solomon et al. 2011), and so there is a need to quantify the ability of models to predict such variability. Indeed, the rapidly increasing global mean temperature rise since the 1970s has slowed and even halted since the 1990s (Easterling and Wehner 2009; Kaufmann et al. 2011; Meehl et al. 2011), which may very likely be a result of natural decadal variability rather than an alteration of secular climate change.

Verification of interannual-to-decadal predictions

Community multi-model experiments such as the series of Coupled Model Intercomparison Projects (CMIPs, Taylor et al. 2012) are providing new approaches for evaluating how model predictions and projections are improving. While the CMIP3 model assessment has previously focused primarily on centennial climate change projections, CMIP5 includes an additional suite of multi-decadal simulations focusing on recent decades out to 2035. The simulations are initialized based on observations—primarily ocean observations. This suite provides an excellent opportunity for using the community hindcast experiments to assess model skill in predicting decadal variability (Goddard et al. 2013). Focused decadal hindcast experiments should use a combination of deterministic and probabilistic metrics to assess hindcast simulations. This approach can be used to (1) test the extent to which initial conditions can provide more accurate predictions of decadal variability compared with uninitialized climate change projections, and (2) address whether a model's ensemble spread is an appropriate representation of forecast uncertainty. However, the main question behind these decadal prediction experiments is whether the ocean initial

conditions provide more accurate predictions than the obtained by the un-initialized climate change projections. Therefore, the baseline for comparison should be whether the initialized predictions are better in some way (i.e., more accurate) than the uninitialized, when compared to the observations for the same period. The question of "What is the baseline the forecast system is trying to beat?" is an important one for any verifying forecasts, especially based on new technology.

The shortness of the observational record and dearth of high-quality climate data (NRC 1998b) makes it difficult to obtain a representative decadal variability record. Initial state knowledge is variable due to nonstationarity in the observing system (Kumar et al. 2012), and climate nonstationarity due to external forcing changes and other processes can lead to decadal forecast biases. As such, it is important that decadal hindcast efforts attempt to address these issues.

Hindcast simulations should focus on those periods in which unexpected climate variability has been observed. As discussed above, the early 21st century is one such period, but could also include periods when basinwide/continental scale climate shifts have appeared in the climate record. Examining such "anomalous" periods will maintain a focus on ocean and atmospheric processes, thus addressing key US CLIVAR goals. Maintaining a process focus, including theoretical studies for understanding the role of ocean dynamics in decadal variability is needed. Beyond hindcast simulations to assess how climate model predictions and projections are improving, the robustness of decadal variability in coupled models needs more investigation, especially with respect to model resolution (Mehta et al. 2011) to allow better separation of natural and forced climate variability.

Statistical model applications

Seasonal snow cover, sea ice, troposphere-stratosphere coupling, and the Arctic Oscillation all exhibit preferred variability on decadal timescales, but the decadal statistics of these climate features need to be better observed. Understanding the coupling between greenhouse gases and the atmosphere as well as other land-ice-ocean forcings with the atmosphere will yield improvements in decadal and centennial climate predictions and projections. Therefore, improvements in simulating these components of the climate system are likely to yield not only greater accuracy on seasonal timescales but may also provide benefit on decadal timescales. For the datasets that are available on a decadal scale, a greater attention on understanding seasonal and regional/hemispheric differences in decadal variability and their causes might shed important light on the controlling processes.

US CLIVAR will facilitate community involvement to determine optimal statistical models and common data formats and to design appropriate model experiments to ensure the best use of statistical models. Statistical models (see section 6.3) may prove to be as useful for benchmarking decadal predictions as they have been for ENSO. Climate models are known to exhibit large deficiencies in simulating land-atmosphere coupling, troposphere-stratosphere coupling, and probably ice-atmosphere coupling. Statistical models that demonstrate superior skill capturing and predicting all these important climate feedbacks should be utilized to provide goals, benchmarks, and even roadmaps for the climate modelers to improve the climate models. Simply improving model resolution, data assimilation, and parameterization schemes is unlikely to yield significant improvement in climate projections without concomitant improvements in our understanding of the physics and coupling between various components of the climate system.

Empirical analysis for quantifying model improvements

Empirical analysis can lead to both empirical models and improvements in the dynamical models. Development of ocean-land-atmosphere-biosphere dynamical models and their coupling needs to continue, with benchmarking against observations. Rigorous testing of models against archived data and data derived from targeted campaigns will quantify model biases and errors. Recent examples include the development of new satellite simulators to test the ability of climate models, and these datasets are now being used to show how cloud representations have improved in the most recent generation of climate models (Klein et al. 2013). Many climate models show large-scale oceanic drifts on decadal timescales after initializations from observations, and the drifts interfere with the decadal prediction efforts. It is very important to identify the key physical processes, assimilation system problems, and observational deficiencies responsible for the drifts and implement better parameterizations to suppress modeled climate drifts and achieve better understanding of decadal predictability.

6.5 Communication of climate research

Climate research is becoming ever more interdisciplinary in nature. Whereas the physical climate system has been the pri-

mary focus of US CLIVAR research activities, it is increasingly clear that interactions between physical, chemical, and biological components are critical not only for understanding climate impacts but also for understanding feedbacks in the system that determine how the physical climate system itself changes. *Interest in the impacts of climate on ecosystems, how the carbon cycle interacts with anthropogenic climate change, and how chemical processes are an important driver of aerosol climate forcing are examples of why US CLIVAR will strive to bridge disciplinary boundaries over the coming fifteen years.*

Effective communication of climate research results, information, and insights is essential for society to reduce vulnerabilities to the impacts of climate variability and change. A well-planned and focused effort to communicate climate science will also allow society to reap the benefits of substantial investments in the climate research enterprise. Anticipation and response to seasonal, interannual, decadal, and longer timescale climate variability offers potentially significant societal benefits.

Communication and dissemination of climate variability science and predictions is an emerging area of emphasis for US CLIVAR, as it is an active and emerging area for the research community. *Research on climate communication is still at a relatively early stage, but some fundamental lessons can be applied to US CLIVAR efforts to facilitate knowledge transfer between the various scientific communities that generate and use information on climate variability and change.*

Knowledge exchange

Multidirectional knowledge exchange between different climate science communities is essential for addressing US CLIVAR goals outlined in Chapter 4. Table 6.1 provides examples of how better communication helps to address each of the goals. Thus, those responsible for generating new observational capabilities and prioritizing the maintenance of the current network can benefit greatly from information from climate models in the form of OSSEs and other means of prioritizing what fundamental climate science is conducted at the process level. Improving the observing network and better selection of process studies will lead to increased understanding of the role of oceans in climate. Effective knowledge exchange also involves communication between the basic and applied science communities and beyond scientists engaging in the physical climate system. The representation of chemical, biogeochemical, carbon cycle, and other ecosystem processes in climate models is becoming more widespread.

US CLIVAR science efforts will benefit from outreach to existing and developing communities of practice and networks of science translators, social science researchers and practitioners, and integrated assessments, such as Cooperative Extension (USDA), RISA (NOAA), Climate Science Centers (Department of Interior), National Weather Service, USGCRP, and the National Climate Assessment. *US CLIVAR will actively seek out and support forums for dialogue, such as needs-assessment workshops, forecast use and evaluation, and developing communities of practice.*

Climate science communication and dissemination

Further studies are needed to ascertain the most appropriate methods and media for the communication of US CLIVAR science, to ensure that CLIVAR science reaches appropriate audiences, and to ensure that climate-relevant science generated beyond the physical climate system is acted on and incorporated into climate models. *Such studies are beyond the scope of the US CLIVAR research agenda; thus, US CLIVAR will benefit from working in partnership with science communities that study and convey climate information to end users. Moreover, US CLIVAR will benefit from efforts to employ research-based insights and best practices in conveying information to end-users.*

Communicating uncertainty

Climate models often fail to inform not because of bad algorithms or bad design, but because of inadequate communication of how they work and what their relative strengths and weaknesses are. Interactions between parameterizations within models, rather than the parameterizations themselves, are increasingly seen as sources of error and need sophisticated collaboration between development teams to untangle and optimize. *US CLIVAR will help achieve better communication, development, evaluation, and exploitation of models by insisting on improvements in the practices of model documentation and comparable quantitative evaluation.*

The use of US CLIVAR science can be improved through the thoughtful articulation and communication of the limitations to the science, uncertainty, and confidence in predictions and projections—using language that is mutually understood by producers and users of climate science information. Studies recommend that effective communication of uncertainty puts uncertainty into context by helping audiences understand what is known with a high degree of confidence and what is relatively poorly understood. End users are sensitive to the language used to convey uncertainty and confidence. Discussing uncertainty with unspecific language can lead to misunderstandings and consequent criticisms (CRED 2009). Verification information is also an essential accompaniment to probabilistic forecasts in order to enable the user to quantify uncertainty based on past model performance (WMO, 2011).

Communicating uncertainty may involve extra effort to understand the institutional, organizational, and cultural contexts of end users, their risk tolerances, and competition with other factors shaping their decision context. US CLIVAR will help provide the information necessary for climate service agencies to develop these efforts. In some cases, providing additional information about forecast skill or ability of models to capture historical climate variations will be needed. Communicating probabilistic information depends on user metrics and familiarity with various ways of communicating forecast success in addition to efforts to educate end users.

Communicating differences among climate variability, anthropogenic forcing, and evolution of the current climate state

Opportunities exist for integration of US CLIVAR science efforts in observations, diagnostic studies, prediction, predictability, and detection and attribution with initiatives to improve the communication and flow of information essential to build capacity to improve end users' comprehension of distinctions between climate variability, anthropogenic forcing, and evolution of the current state of climate. In practice, this will amount to a process of continual adjustment as knowledge advances, user needs change, and understandings of the problem evolve (SPARC 2010). Studies suggest that regular communication with end-user audiences and efforts to co-produce science and knowledge products with science communication intermediaries and end users will foster improved understanding (Lowrey et al. 2009). Interpretation of seasonal climate outlooks along with GCM-based and scenario-based climate require comparison with observations; thus, access to high-quality historical databases and materials that make the linkages between historical climate variations, paleoclimate, future projections, and impacts will also be necessary (Dow and Carbone, 2007).

Climate science communication best practices

Research on best practices for climate science communication is essential for promoting informed use of climate science and for building trust and credibility in climate science insights, predictions, and projections. This relatively new area of research is rapidly changing in the face of new technologies and media for conveying information. Strategies for communicating climate science, regardless of the medium, rest on defining the desired end-user audiences for particular climate information, understanding the mental models of those audiences, and their attitudes regarding particular frames for information (CRED 2009). Framing information in a way that will convey both salience and credibility may require establishing the linkages between basic climate science insights, regional forecasts, local impacts, costs or threats, and potential actions. In US CLIVAR the main audience may be the broad scientific community, but that may not be the only audience. US CLIVAR can take advantage of gained knowledge on framing its science in its own communications, e.g., through the newsletter and the website.

Building links with research communities

Fostering the development of a network of climate science communities, infrastructure, and institutions is needed to effectively and efficiently achieve US CLIVAR goals. *A well-planned engagement strategy will help build upon US CLIVAR efforts such as the Postdocs Applying Climate Expertise (PACE) fellowship program. US CLIVAR will foster connections with other scientific communities such as Ocean Carbon and Biogeochemistry to address how the ocean will respond to increasing atmospheric carbon dioxide.* The role of the oceans in sea ice variability and change is poorly understood and will require better linkages with CliC. IGAC and SOLAS play important roles in understanding the factors controlling anthropogenic and natural aerosols. Because both play important roles in determining the regional expression of decadal and multidecadal climate changes, especially via their impacts on clouds, *US CLIVAR will encourage better collaboration with these communities to ensure that the benefits of our improved understanding translate into improved representation of aerosol-cloud interactions in climate models.* US CLIVAR will build links with other research communities through their professional societies, e.g., American Society of Civil Engineers, American Planning Association, Ecological Society of America, and consider the priorities identified by these other communities in CLIVAR's own planning. US CLIVAR will engage these communities through its own working groups and workshops, by working with agency operational centers, agencies responsible for projecting the status of resources influenced by climate, and professional societies representing them, and by encouraging agencies to work across disciplinary boundaries.

Chapter 7

Management and Implementation Activities

US CLIVAR is a multiagency-sponsored climate research program that involves the participation of many volunteer scientists, typically 100–200 at any given time. This chapter describes the program's management structure and implementation activities that have proven successful for achieving the program's goals during its first 15 years.

7.1 Management

US CLIVAR is managed by a three-part structure designed to facilitate close collaboration between the climate science community and the funding agencies that sponsor climate research as shown in Figure 7.1. The management structure consists of: (1) the scientists, who populate the Scientific Steering Committee (SSC) and three panels of research community members to plan the science program and its ongoing implementation; (2) an Inter-Agency Group (IAG) of program managers who fund the research and planning efforts; and (3) a Project Office funded by the IAG and overseen by the SSC to provide operational support for the program.

Scientific Steering Committee

The US CLIVAR Scientific Steering Committee (SSC) provides overall scientific and programmatic guidance for the program, developing the science plans and implementation strategies to ensure that US CLIVAR progresses towards achieving its science goals. The SSC identifies science gaps, selects research themes, promotes balance among program elements, ensures coordination with international CLIVAR and USGCRP elements, keeps the NRC apprised of program status, comments on agency implementation, and provides oversight and guidance to the US CLIVAR Project Office.

Membership is comprised of nine community leaders with expertise spanning the broad science goals of the program. Included are three Executive Committee members (a chair and two co-chairs) and two co-chairs for each of the three panels, ensuring connection and communication among the panels and the program leadership. Executive Committee members are selected by the IAG. They serve for three years with staggered rotation to ensure continuity as membership evolves. The other six SSC members, the co-chairs for each of the three panels, are determined by the SSC based on nominations from the current panel co-chairs. Panel co-chairs serve on the SSC for a 2-3 year period. Their terms are staggered as well to ensure continuity in leadership of the panels.

The SSC convenes a US CLIVAR Summit, currently held annually, enabling direct communication among the SSC, its panels, the IAG, the Program Office, and invited guests. These Summits provide a critical pathway for exchanging information on progress of implementation activities, US research challenges (formerly themes), International CLIVAR activities, and funding agency interest. They generate a list of specific action items to guide research planning and implementation activities in the upcoming year. Between Summits, the SSC holds an in-person meeting with the IAG to maintain an ongoing discussion. Much of the work of the SSC and its engagement of the Project Office are conducted via teleconferences through the year.

Panels

The SSC establishes panels to coordinate community input and organize implementation in specific areas of research. From 2005 to the present (2013), US CLIVAR has operated with the following three panels:

The **Phenomena Observations and Synthesis (POS) Panel** to improve understanding of climate variations in the past, present, and future, and to develop syntheses of critical climate parameters while sustaining and improving the global climate observing system.

The **Process Study Model Improvement (PSMI) Panel** to reduce uncertainties in the general circulation models used for climate variability prediction and climate change projections through an improved understanding and representation of the physical processes governing climate and its variation.

The **Predictability, Predictions and Applications Interface (PPAI) Panel** to foster improved practices in the provision, validation and uses of climate information and forecasts through coordinated participation within the US and international climate science and applications communities.

The topics of the panels align with the five science goals articulated in Chapter 4, thereby enabling the panels to readily engage in planning implementation activities to achieve the goals.

Each panel is comprised of up to 12 experts on topics for which the panel is responsible. Care is given in composing the membership of each panel to ensure balance among disciplinary areas of research, institutional affiliation, and diversity. Panel members serve for a three-year term with the opportunity for a second term. When members rotate off, the panel co-chairs identify the specific expertise needed to fill out the membership, reflecting ongoing and new research directions of the program. An annual search for new panel members proceeds through a public call for nominations administered by the Project Office. Nominees are reviewed and new members are selected by the SSC.

Terms of reference, established by the SSC for each panel, charge the panels to: advise US CLIVAR on research priorities, identify research gaps, and develop suitable milestones that promote funding opportunities; develop and encourage activities (e.g., community workshops, commissioned studies, Working Groups) that further develop and implement US CLIVAR goals, engage in cross-cutting strategies, and address research challenges; advise on the adequacy and effectiveness of Working Group plans and implementation; consider necessary coordination with other national and international activities to develop integrated, efficient, and effective plans; liaise with other US CLIVAR panels to ensure relevant needs are considered in their

efforts; and generate a list of accomplishments and progress over the past year, action items for the panel, and set of recommendations for SSC and funding agency consideration.

The three panels convene in person at US CLIVAR Summits. They breakout into separate business sessions to review progress of ongoing activities, survey emerging opportunities, and identify specific action items and recommendations for SSC consideration, provided in the closing plenary session. Panels often meet jointly during the Summit for briefing and discussion on topics of mutual interest and responsibility. In the interval between Summits, the panels continue to meet through periodic teleconferences to progress on action items and consider new business that may arise. Panel co-chair membership on the SSC ensures ongoing two-way communication between the panels and their parent body.

Interagency Group

The US CLIVAR science goals support multiple funding agency missions and thereby motivate collaborative multi-agency sponsorship. Agency program managers voluntarily convene as an IAG to coordinate the implementation of research activities in support of US CLIVAR goals. The agency programs managed by members of the IAG fund the individual research projects that collectively comprise the US CLIVAR program. The membership draws from the National Aeronautics and Space Administration (NASA), the National Science Foundation (NSF), the National Oceanic and Atmospheric Administration (NOAA), the Department of Energy (DoE), and the Office of Naval Research (ONR). Members meet monthly with participation of the Program Office to: confirm agency interests, announcements, and budgets; review plans and progress of the SSC, its panels, and working groups; identify and discuss research opportunities of mutual interest; develop and implement joint research proposal solicitations; initiate and foster Science Teams and Working Groups; and review requests for supporting workshops and science meetings, funding those of greatest relevance and potential return. Program managers from other programs and agencies are welcome to join IAG discussions. In addition to participating in Summits, IAG members also actively engage the SSC at its in-person meetings. Members of the IAG whose programs contribute to funding of the Project Office annually approve the scope of work and funding of the Project Office, including contributions to International CLIVAR and its Project Office.

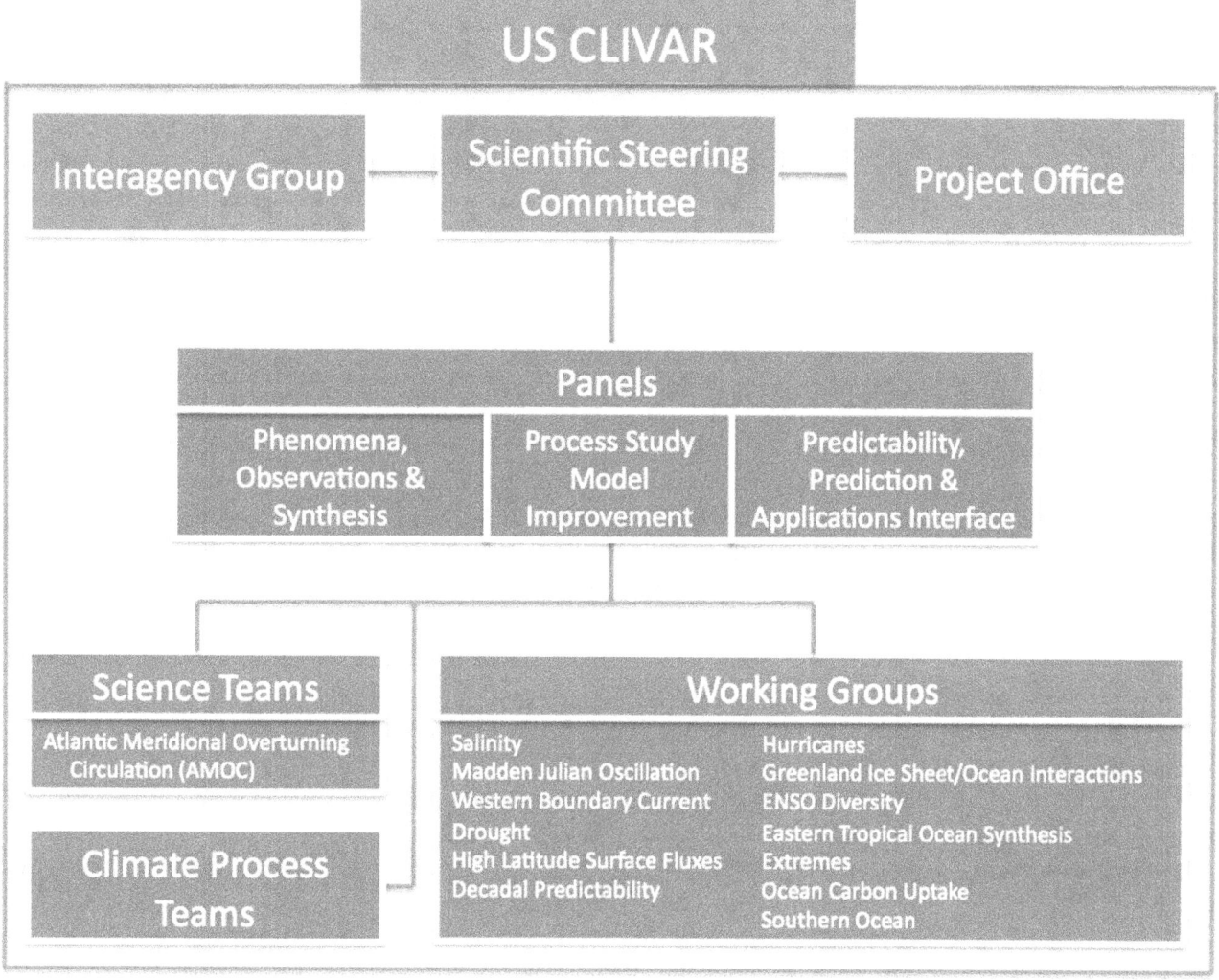

Figure 7.1 *US CLIVAR organizational structure*

Project Office

The Project Office is responsible for ensuring all scientific and programmatic coordination and implementation is completed as guided by the US CLIVAR SSC and supported by IAG. The Office supports the ongoing day-to-day activities of the program, which include: arranging and supporting meetings of the IAG, SSC, and panels; organizing summits, workshops, meetings and agency briefings; assisting Science Teams to coordinate planning and reporting of results; coordinating the solicitation, review, and support for Working Groups; facilitating US scientific engagement of International CLIVAR and organizational contributions to its Project Office; engaging other programs to explore collaborative interests; and overseeing administrative matters such as budget planning/execution and progress reporting. The Project Office plays a central role in facilitating

communication across the program, providing updates among the SSC, panels, and IAG, and conveying US CLIVAR research and program developments to the broader research community through websites, reports, newsletters, monthly news-gram, and twitter feeds.

7.2 Implementation activities

US CLIVAR will continue to employ the range of implementation activities that have already proven successful in coordinating research and accelerating progress toward program goals. Most of the strategies involve groups of experts who survey current state of knowledge, identify gaps, and recommend specific activities to address the gaps in knowledge. Some strategies have been developed by the US CLIVAR SSC (e.g., Climate

Process Teams), while others are borrowed and modified from other programs (e.g., Science Teams).

Climate Process Teams

As described in Chapters 2 and 6, Climate Process Teams (CPTs) are intended to: (1) quantify and reduce uncertainties associated with specific processes, realizing these advances within climate models; and (2) recommend additional process and observational studies needed to study the processes in question. Teams are composed of observationalists, process modelers, and parameterization developers working collaboratively with climate-model developers at US modeling centers. They are designed to provide a responsive two-way link between process-oriented research (e.g., short-duration observation campaigns, process parameterizations) and climate model development, thereby accelerating the transfer of process research findings into parameterizations leading to improved climate simulations. The CPTs bridge all US CLIVAR goals.

Past and present CPTs typically involve many young scientists and students. The training through participation received by these young scientists is multidisciplinary, and helps them learn to frame their future research goals toward integration of process studies and climate models. The CPTs live on after their completion through this training process.

Solicitation, review, and award of CPT projects are coordinated among participating agency programs. Solicitations typically require the participation of the modeling centers of each sponsoring agency. Awarded projects report annually on progress and challenges to the US CLIVAR Process Study Model Improvement Panel.

Working Groups

US CLIVAR Working Groups are limited-lifetime, action-oriented groups of scientists, typically with 8–12 core members. They are assembled to coordinate and implement focused activities for the benefit of the broader scientific community. A major aim of the Working Groups is to expedite coordinated efforts towards specific scientific activities and objectives, for example: assessing existing or developing new data and modeling products and capabilities, leading community-wide analyses or syntheses of current state of understanding, and/or developing scientific and implementation recommendations on specific subjects for further consideration by US CLIVAR panels. Furthermore, Working Groups engage community-wide participation whenever possible through open workshops, web pages, newsletters, journal articles, and reports. Some serve to facilitate interdisciplinary research activities, forging collaboration among scientific communities. Working Groups are intended to have an impact that extends beyond the conclusion of their funded lifetimes; each galvanizing a *community* of researchers to further advance scientific exploration.

New Working Groups are solicited through an open announcement process inviting submission of prospectuses outlining activities, milestones, support needs, and anticipated benefits to the CLIVAR program and the climate research community. To promote grassroots engagement, the announcement does not typically specify science topics. Each prospectus presents a list of potential working group members for consideration. SSC review of the submitted prospectuses is considered by the IAG in determining which will be initiated.

When funded, the Working Group chairs immediately establish the membership with approval of the SSC, and engage the Program Office to launch the effort through teleconference meetings. Working Groups routinely report on progress at Summits and provide updates to the SSC, panels, IAG, and other groups as appropriate.

Science Teams

Science Teams differ from Working Groups in that they are coordinated more directly by the funding agencies. Extending an approach originating with NASA, US CLIVAR Science Teams are a multi-agency funded group of Principal Investigators working on a common research topic of mutual and long-term interest of participating agencies. A Science Team is assembled to foster communication and collaboration among individual research projects. The Team operates to define, refine, and track progress toward science goals, to promote research within the broader science community, and to serve as a US focal point to further inter-program and international scientific coordination. To achieve these objectives, the Science Team establishes an implementation plan outlining near-term priorities, research tasks, and specific activities to be undertaken. It organizes PI meetings and special sessions at scientific conferences to survey recent research advances, identify science gaps, and refine priorities for accelerating progress. Teams are encouraged to produce annual reports synthesizing collective progress of projects toward achieving collective objectives, organize workshops to focus planning on specific science topics, and publish review articles and dedicated journal volumes to share results with the broader community. Periodic external reviews of Teams will be employed to evaluate the adequacy of science plans, successes and impediments encountered in implementation, and the degree to which synthesis is achieved beyond individual research projects.

The US Atlantic Meridional Overturning Circulation (AMOC) Science Team is the first such team established by US CLIVAR in 2008. Participating agency programs in NASA, NOAA, NSF, and DoE have identified relevant existing projects as contributing to the US AMOC Program and the project PIs as members of the Science Team. The US AMOC Science Team has been charged with responsibility to establish an implementation plan and accomplish its objectives with guidance and oversight from the supporting agencies. The Team has self-organized into four Task Teams to develop specific research tasks and near-term priorities spanning (1) observing system implementation and evaluation, (2) assessment of AMOC state, variability, and change, (3) investigation of variability mechanisms and predictability, and (4) evaluation of global climate and ecosystem impacts.

Additional Science Teams are expected to be employed to organize and implement community participation for a range of US CLIVAR science topics, each with a critical mass of funded research projects among interested agencies. Consideration and decision to establish a new Science Team is made by two or more funding agencies with shared interest in a specific US CLIVAR research priority topic for which the agencies determine there are clear advantages for progress through collaboration and synthesis among individual funded research projects. Members are identified and confirmed by agency program managers. The expected duration of a Science Team is five to ten years.

Meetings and workshops

US CLIVAR sponsors meetings (e.g., US CLIVAR Summit, Science Team meetings) and workshops (e.g., organized by Working Groups) of members of the science community to survey research needs and identify specific implementation approaches and opportunities that advance US CLIVAR goals.

Requests for scientific workshop and meetings are considered twice a year by the IAG. Guidelines for seeking US CLIVAR sponsorship request submission of formal requesting outlining the nature, scope, objectives, relationship to other meetings, potential participants, deliverables, and statement of benefits to US CLIVAR. Priority is given to those efforts that demonstrate high relevance and payoff for US CLIVAR investment.

Opportunities for early career investigators

Early career scientist participation in organizing and implementing US CLIVAR research is essential to the success of the research agenda, particularly for a program that extends over decades. Mechanisms for entraining next generation researchers into the field need emphasis, even in restricted budgetary environments. Research opportunities for young investigators are promoted through early career scientist solicitations, postdoc program solicitations, and student training programs (e.g., NCAR Advanced Studies Program). Whenever feasible, travel grants and reduced registration fees for early career scientists and students will be employed in US CLIVAR-sponsored meetings and workshops to promote participation of young scientists in science reviews and implementation activities. US CLIVAR panels, working groups, and workshop organizing committees will promote membership opportunities for early career scientists and will entrain young scientists as members. The CPTs, Science Teams, and Working Groups are all encouraged by US CLIVAR to utilize and help train young scientists where possible.

Agency solicitations and project awards

Agency solicitations inviting and awarding research projects is a primary mechanism for implementing US CLIVAR research. It is through individual research projects, often coordinated collectively, leveraged, or synthesized through the above-mentioned implementation strategies that the science of US CLIVAR advances.

Agencies may employ joint solicitations to collectively invite, review, select, and award projects on a focused US CLIVAR topics, thereby improving coordination among agency-sponsored research and broadening the opportunity for participation by the community. Such solicitations have been utilized for process studies including field campaigns, model simulation intercomparison and evaluation studies (e.g., Climate Model Evaluation Project), and diagnostic and synthesis analyses of datasets (e.g., Drought in Coupled Models Project). The use of rapid, small research project solicitations can be employed to incrementally enhance ongoing projects, leveraging them to address an additional objective or to accelerate progress.

Chapter 8
Program Cooperation and Coordination

US CLIVAR seeks active engagement with other Earth-science communities, both within the US and internationally. This engagement will foster activities that address shared science questions at the interface of traditional disciplinary program boundaries. The collaborations that are envisioned target important areas of infrastructure (observing systems, data centers, research platforms, modeling and prediction centers, and national and international scientific assessments) to which US CLIVAR contributes and upon which the program relies.

8.1 US Global Change Research Program

The USGCRP coordinates and integrates federal global change research across thirteen participating agencies, including the five represented on the US CLIVAR Interagency Group. US CLIVAR provides the program contribution to the USGCRP on understanding the ocean's role in climate variability and predictability.

All US agency-designated US CLIVAR research projects, including the US agency sponsorship of the US and International CLIVAR Project Offices, are included in the US Global Change Research Program. The US CLIVAR Project Office serves as a primary contact point for the National Coordination Office (NCO) of the USGCRP for the topic of climate variability and predictability research. The Project Office furnishes updates on CLIVAR-related research accomplishments and plans to the NCO for inclusion in the USGCRP Our Changing Planet annual report to Congress. The Project Office also works with the NCO to arrange briefings on US CLIVAR for the Subcommittee on Global Change Research, the interagency body responsible for steering the USGCRP.

The USGCRP recently released a new ten-year strategic plan (USGCRP, 2012) outlining four strategic goals: Advance Science, Inform Decisions, Conduct Sustained Assessments, and Communicate and Educate. US CLIVAR contributes most readily to the first goal to advance scientific knowledge of the integrated natural and human components of the Earth System. The US CLIVAR cross-cutting strategy for communicating climate variability outlined in Chapter 6 promotes links to the other USGCRP goals: providing the scientific basis to inform and enable timely decisions on adaptation and mitigation; build sustained assessment capacity that improves the Nation's ability to understand, anticipate, and respond to global change impacts and vulnerabilities; and advance communications and education to broaden public understanding of global change and develop the scientific workforce of the future. As USGCRP interagency working groups are organized to address these goals, US CLIVAR will continue to cooperate and communicate with research programs within the US on interdisciplinary questions and challenges of climate variability and change.

Interdisciplinary research needs outlined for the research challenges in Chapter 5 will be addressed through coordination with other US research programs of the USGCRP as identified in the following sections.

Land-surface hydrology and terrestrial ecosystem impacts research

Representation of land-surface processes and their interactions with ocean, atmosphere and ice processes must be represented in climate models to achieve accurate representation and predictions of the climate system. Understanding the relative roles and interactions of the ocean-atmosphere coupling with land-atmosphere coupling has motivated cooperative activities with the land-surface community for over a decade. Collaborations with US Regional Hydroclimate Projects and the North American Monsoon Experiment have investigated the coupled ocean-land-atmosphere system and the predictability of the summer climate and hydrologic cycle over North America. Continued cooperative efforts with the land-surface research community in the US are desired.

For the research challenge climate extremes, US CLIVAR seeks collaboration with land-surface hydrology programs such as the emerging North American Water Program, with its focus on analyzing variations, trends, and extremes in the water cycle over North America.

The impact of climate variability and change on terrestrial ecosystems is a future possible focus of the program. Land biological impacts and feedbacks have been addressed in part through the US CLIVAR drought and extremes foci, and future work would also likely engage US land-surface hydrology programs and the USGS Climate Science Centers, which are identifying needs for climate science within DOI and its Landscape Conservation Cooperatives.

Carbon cycle, ocean biogeochemistry, and marine ecosystem research

Understanding how climate will respond to future atmospheric concentrations of CO_2 and other carbon-containing greenhouse gases, and how carbon sources and sinks will change in response to a changing climate, are questions shared by US CLIVAR and the US Carbon Cycle Science Program. Since 2003, the US Global Ocean Carbon and Repeat Hydrography Program has carried out the systematic and global re-occupation of select hydrographic sections to quantify changes in storage and transport of heat, fresh water, and carbon dioxide, thereby supporting the objectives of US CLIVAR and the US Carbon Cycle Program. Further joint program exploration of the coupled physical-biogeochemical processes and feedbacks and their representation in coupled Earth system models are needed to explore the future state of the climate, carbon sources and sinks, and related ecosystem response. The programs have embarked on scoping and implementing collaborative research efforts to be undertaken over the next decade.

Beginning in 2009, US CLIVAR initiated a collaboration to address this challenge with the US Ocean Carbon and Biogeochem-

istry (OCB) Program, the mission of which is to study the impact of oceanic variability in the global carbon cycle, in the face of environmental variability and change through studies of marine biogeochemical cycles and associated ecosystems. Two working groups jointly sponsored by US CLIVAR and OCB are underway to develop observation-based metrics to evaluate modeling of processes and feedbacks that govern (1) the uptake of ocean carbon in CMIP5 models and (2) the heat and carbon uptake in model simulations of the Southern Ocean. Analyses will help guide future observational campaigns and motivate model improvements.

Interactions between climate and marine ecosystems may be addressed not only through a future expansion of joint activities with OCB but also with other groups. For example, preliminary work on AMOC impacts on marine biology are developing. US CLIVAR can also work through International CLIVAR and its new research opportunity on marine biophysical interactions and upwelling systems. Partnerships can be explored with NOAA Coastal Ocean and National Marine Fisheries Science Centers. Another program collaboration could be explored with the Comparative Analysis of the Marine Ecosystem Organization (CAMEO), a joint NSF-NOAA research program initiated in 2009 to provide an understanding of and predictive capability for marine ecosystem organization and production through examination of how climate variability and fishing pressure affect them. US contributions to the International Integrated Marine Biogeochemistry and Ecosystem Research (IMBER), particularly the regional Sustained Indian Ocean Biogeochemistry and Ecosystem Research (SIBER) and the Integrated Climate and Ecosystem Dynamics (ICED) study of circumpolar analyses of Southern Ocean climate and ecosystem dynamics, offer yet another potential interface for US CLIVAR to develop.

Atmospheric aerosol-cloud interactions

As stated in the Chapters 4, 5, and 6, there is suggestion that a portion of decadal variability is driven by changes in anthropogenic aerosols. Yet the sources and fate of aerosols and the cloud microphysical processes that control clouds and aerosols are not well represented in climate models. US CLIVAR has joined with the aerosol and cloud research communities to mount process studies (e.g., VOCALS) and has sponsored model development efforts (e.g., CPTs) to improve understanding of aerosol and cloud processes and their representation in climate models. Future collaborations with the US community, International IGAC, SOLAS, and GEWEX programs are sought to further improve the understanding and modeling of aerosol-cloud interactions.

Polar and cryospheric research

To address the polar-climate research challenge outlined in Chapter 5, cross-disciplinary collaborations involving oceanographers, atmospheric scientists, glaciologists, and land-surface hydrologists are needed to scope research strategies, mount observational campaigns, understand coupled ocean-ice-atmosphere-land processes, and improve coupled model performance in high latitudes. Such collaboration was undertaken by the US CLIVAR Working Group on High Latitude Surface Fluxes, which assessed the status of flux products for momentum and heat in high-latitude regimes, evaluated commonalities between the Arctic and Antarctic regions, and identified priorities for continued and new *in situ* flux observations, improved satellite flux observing capabilities, more accessible observations and flux products, and flux intercomparisons in polar regions (Bourassa et al. 2013).

US CLIVAR is engaging the Study of Environmental Arctic Change (SEARCH) Program to define common research topics, including changes in climate, sea ice extent, and ice sheet mass in the Arctic basin, and the impacts on ocean circulation and regional sea level. The US CLIVAR Working Group on Greenland Ice Sheet-Ocean Interactions has summarized the current state of knowledge and identified key physical aspects of Greenland's coupled ice-sheet/ocean/atmosphere system (Straneo et al. 2013). The effort may provide a foundation for future collaborative research to be undertaken in partnership with SEARCH.

As mentioned previously, US CLIVAR is also investigating processes in the Southern Ocean through its Working Group on Southern Ocean Heat and Carbon Uptake. The Working Group aims to improve understanding of the role of mesoscale eddies in the heat and carbon uptake by the Southern Ocean and improve understanding of how the Southern Ocean stratification, circulation, and heat and carbon uptake will respond to a changing climate. The effort will identify critical observational targets to inform implementation of the Southern Ocean Observing System (SOOS), and will select observation-derived metrics to evaluate coupled climate model simulations, with an aim of identifying model deficiencies in representing critical processes. US CLIVAR seeks to partner with the US Antarctic Research Program to collaborate on future research opportunities for understanding the Southern Ocean and Antarctic roles in and responses to climate variability and change.

8.2 The World Climate Research Program

The influence of the global ocean on the global climate system inherently requires global-scale observations, analyses products, process understanding, modeling, and synthesis. US science interests intersect with those of other countries, such that collaborative research program planning and coordinated implementation can provide shared benefits, beyond those capable of being achieved by working independently. The following sections describe the international linkages of the US program, underscoring the benefits of coordinating multi-country commitments to shared priorities and of contributions to capacity building.

US CLIVAR is the US contribution to International CLIVAR, one of four core programs of the World Climate Research Program (WCRP). Cooperation and collaboration with International CLIVAR occurs on an ongoing basis as described below. Collaboration of the US program with the WCRP and its other three programs is principally through the International CLIVAR interface. On occasion, US CLIVAR dialogues with the other international programs on specific topics in which US CLIVAR is developing activities in conjunction with or on behalf of International CLIVAR. Examples of these topics are also provided here.

International CLIVAR

International CLIVAR, a core program of the WCRP, provides the forum for collective multi-country planning and implementation of research on the role of the ocean in climate variability and predictability. Launched in 1995, International CLIVAR's mission is to improve understanding and prediction of ocean-atmosphere interactions and their influence on climate variability and change, to the benefit of society and the environment (WCRP, 2012).

Now approaching its 20th year, International CLIVAR is evolving its structure and identifying research priorities for the next decade building upon the successes of its observation, modeling, and regional/basin panels. To address its newly stated mission, International CLIVAR has articulated seven initial focused and integrated research opportunities:

- intraseasonal to interannual variability and predictability of monsoon systems;
- decadal variability and prediction of ocean and climate variability;
- trends, nonlinearities and extreme events;

- marine biophysical interactions and dynamics of upwelling systems;
- dynamics of regional sea level variability;
- ENSO in a changing climate; and
- planetary heat balance and ocean heat storage.

These research opportunities are areas of critical importance and primed for progress through international coordination. The goal is for CLIVAR to develop a framework/program structure that is flexible and can respond to the needs of the CLIVAR community. The research opportunities will be advanced through new, targeted coordinated activities and the continued development of the core CLIVAR cross-cutting capabilities.

The mission, objectives, and science foci above align with those of the US program presented in Chapters 2–5, albeit with somewhat different structure. US CLIVAR is thus well poised to contribute to international program objectives and link with efforts of other countries doing the same. US CLIVAR indeed evolves in the context of the international program, being both influenced by new and emerging research directions internationally, and influencing the directions of the international program. US members of the International CLIVAR Scientific Steering Group, panels, and working groups, many of whom also serve on US CLIVAR's SSC, panels, teams, and working groups, share updates of US program interests and activities to inform the international planning effort as well as feedback to the US program, the priority foci and interests of the international research community. The complementarity of the international and US program plans reflects this ongoing dialogue and collaboration, and it is expected that US CLIVAR will implement research projects that are collaboratively sponsored across countries.

Global Energy and Water Exchanges (GEWEX)
US CLIVAR, working in tandem with International CLIVAR, engages other core projects within the WCRP to collaborate on science topics of mutual interest, most readily the Global Energy and Water Exchanges (GEWEX) Project with its focus on studying the dynamics and thermodynamics of the atmosphere and interactions with the Earth's surface. CLIVAR science exploits GEWEX efforts to provide the precise global estimates of surface radiation fluxes and heating needed to predict transient climate variations and decadal-to-centennial climate trends. GEWEX efforts to evaluate the role of evaporation and precipitation processes in regional rainfall anomalies intersects well with US CLIVAR's interest in understanding and predicting

climate variations, including drought, and extremes in precipitation and temperature on a range of time and space scales. The programs also share interests in improving process understanding and its translation to improved model representation of atmospheric radiation, clouds, and land-surface interactions (e.g., evapotranspiration, soil moisture). Fruitful future collaborations will continue to be sought with International GEWEX and US projects focused on a range of topics, including process understanding (water exchanges, ocean/land surface/atmosphere interactions), observational and model datasets, analyses and reanalyses, climate and Earth system modeling, and predictability/prediction.

Climate and Cryosphere (CliC)
Links with WCRP's CliC Program should emerge through the US CLIVAR research challenge on polar climate, with shared questions regarding sea and land ice extent and changes, the ocean's role in them, and the impacts of changing ice on climate and the ocean. Discussions in the US of these science topics for the Arctic region are underway between the US CLIVAR and the SEARCH program. Joint activities in the Arctic region in the US, developed through such collaboration will inform International CLIVAR and CliC dialogue and joint planning efforts. For the southern polar region, the US CLIVAR Southern Ocean Working Group's focus on improving simulation of the Southern Ocean in climate models includes cryosphere-related topics such as observations of heat transport to and under Antarctic ice shelves, observation-based metrics for sea ice, and model biases related to iceberg representation. Interactions with CliC on these and other southern polar topics can be pursued, particularly through the International Southern Ocean Panel cosponsored by CLIVAR, CliC, and the Scientific Committee on Antarctic Research (SCAR).

Stratospheric Processes and their Role in Climate (SPARC)
The WCRP SPARC Program's theme on climate variability and change intersecting with CLIVAR may have implications for extremes and also decadal variability. Areas of common interest include dynamic stratosphere-troposphere interactions, solar influences on the stratosphere and upper troposphere, and changes in stratosphere chemistry that impact tropospheric circulations and atmospheric temperature trends. Collaborative links with SPARC will be pursued through the International CLIVAR interface.

Other International Programs

The US and international program connections identified above are not wholly inclusive of all potential program interactions. Collaborations with other programs with similar goals are welcomed. Interested parties may contact the US CLIVAR Project Office to express interest in collaborating.

8.3 Enabling infrastructure

It is important to recognize that US CLIVAR research is enabled by critical infrastructure investments organized and funded under other auspices. Examples of such infrastructure, some of which are established through the work of CLIVAR (e.g., observing and modeling systems), are provided below. As a customer of these systems, US CLIVAR needs to continuously engage and inform their sponsors of the requirements and uses of each system for climate research.

Sustained in-situ and satellite observing systems

US CLIVAR plays a central role in motivating and utilizing in-situ and satellite ocean and atmosphere observing systems to monitor, understand and model climate variability. Key observing systems beneficial to US CLIVAR research include:

- The **Global Tropical Moored Bouy Array** of over 110 fixed moorings spanning the tropical Pacific (TAO/TRITON), Atlantic (PIRATA), and Indian Ocean (RAMA) basins instrumented to measure both upper ocean and surface meteorological variables involved in ocean-atmosphere interactions, providing knowledge of critical processes important to monitoring, understanding, and prediction of ENSO and PDO in the Pacific, the meridional gradient mode and equatorial warm events in the Atlantic, the Indian Ocean Dipole and intraseasonal Madden-Julian Oscillation in the Indian Ocean, the mean seasonal cycle and monsoon circulation, and decadal trends that may be related to global warming.

- **Ocean Reference Stations**, a global array of 60 moored surface and subsurface buoys instrumented to provide the most accurate possible long-term climate data records of heat, momentum, freshwater, and gas (e.g., CO_2) exchanges across the air-sea interface, ocean currents and transport to a depth of 5000m, and biogeochemical properties within the water column in key ocean regimes (e.g., narrow Western Boundary Currents, trade-wind sites) to detect sudden changes and events, to calibrate

remotely sensed measurements, to elucidate climatically sensitive processes, and to evaluate and improve models.

- The **Global Expendable Bathythermograph (XBT) Network** comprised of over 30 fixed transects spanning all ocean basins along which measurements of water temperature from the surface to 850m are made every 25km and at least 4 times a year (for high-density transects), resolving both the oceanic boundary currents and the corresponding interior heat and mass circulations of the global oceans.

- The **US Global Ocean Carbon and Repeat Hydrography Program** and the international Global Ocean Ship-Based Hydrographic Investigations Program (GO-SHIP) providing systematic and global re-occupation of select World Ocean Circulation Experiment (WOCE)/Joint Global Ocean Flux Study (JGOFS) hydrographic sections (currently 46 lines globally) that measure pressure, temperature, salinity, carbon, nitrogen, oxygen, and other biogeochemistry variables from the surface to the deep ocean, providing the ability to quantify full depth changes in storage and transport of heat, fresh water, and CO_2 globally on decadal timescales.

- The global **Argo Array** of 3,000 free-drifting profiling floats that measure temperature, salinity, and velocity of the upper 2000m of the ocean, providing a quantitative description of the changing state of the upper ocean, sea level, and the patterns of ocean climate variability from months to decades, including heat and freshwater storage and transport.

- The **Global Drifter Program** array of 1250 satellite-tracked free-drifting surface drifting buoys that measure mixed layer currents, SST, atmospheric pressure, winds and salinity, supporting climate monitoring, research, seasonal-to-interannual predictions, and satellite SST measurement calibration.

- Satellite data products crucial for monitoring and studying climate variability and change and for improving the descriptive and predictive skill of climate models, e.g., the **TOPEX/Poseidon** altimeter and **Jason-1** (sea level); infrared and microwave channels (SST), the NASA Quick Scatterometer (**QuikSCAT**; ocean surface vector winds); the Gravity Recovery and Climate Experiment (**GRACE**; spatial mass changes), **Aquarius** (salinity), passive radiometers, such as Special Sensor Microwave Imager (**SSM/I**) and Advanced Microwave Scanning Radiometer (**AMSR**; sea ice), the Ice, Cloud, and land

Elevation Satellite laser altimeter (**ICESat**: ice sheet topography and sea ice freeboard), and the Constellation Observing System for Meteorology, Ionosphere, and Climate (**COSMIC**; precipitable water vapor), in addition to measurements of total solar irradiance, stratospheric aerosols, and land-cover and land-use change.

The coordination and oversight for these observing systems are largely provided under international auspices, i.e., through the Global Ocean Observing System (GOOS), the Global Climate Observing System (GCOS), the Committee on Earth Observation Satellites (CEOS), and the Joint Technical Commission for Oceanography and Marine Meteorology (JCOMM). US and International CLIVAR, with their common focus to help identify requirements and inform design of a sustained climate observing system, engage these programs that share responsibility to implement, maintain, and update the systems. The Global Synthesis and Observations Panel (GSOP) of International CLIVAR has an important role to play in this regard. US CLIVAR will continue its strong record of applying existing or new technologies in the context of field experiments and pilot observation programs that help to evaluate the contribution of new observations and whether they should be sustained and expanded. US contributions will aim to aid the Ocean Observations Panel for Climate (OOPC) in its responsibilities to evaluate and evolve the ocean observing system.

Data centers

US agencies sponsor a distributed set of data centers with responsibilities for collecting, organizing, and serving climate data and related metadata essential for research on climate variability and predictability.

- NOAA supports three data centers: the National Climatic Data Center (NCDC) providing *in situ* atmospheric, land and marine surface, NOAA satellite, paleoclimate, and NOAA climate model data; the National Ocean Data Center (NODC) providing *in situ* and satellite oceanographic data; and the National Geophysical Data Center (NGDC) providing paleo-oceanographic, snow, and ice data.
- NASA supports a system of twelve Distributed Active Archive Centers (DAACs) providing oceanographic, atmospheric, terrestrial, cryospheric, biogeochemistry, solar, and human dimensions data from NASA's past and current Earth-observing satellites and field measurement

programs, as well as NASA climate model data.
- NSF supports the NCAR data services of the Earth Observing Laboratory (EOL) data archive providing atmospheric, oceanographic, and other geophysical datasets from operational sources, scientific research programs, and process studies; the NCAR High Performance Storage System providing NCAR Coupled Earth System Model (CESM) data; and the DAAC for COSMIC data.
- DoE supports the Program for Climate Model Diagnosis and Intercomparison (PCMDI) and its leadership role in the Earth System Grid Federation providing storage and distribution of multiple Coupled Model Intercomparison Project (CMIP) simulations and the pilot DoE/NASA Obs4MIPs assembly of observational datasets for model validation.

In addition to the agency data centers, specialized data centers have been established for many of the in-situ observing systems listed in the preceding section (e.g., the CLIVAR & Carbon Hydrographic Data Office, Argo Global Data Assembly Centers),

The data centers perform data rescue and digitization of historic records, conduct quality control of datasets, establish formatting protocols and metadata requirements, and provide access, analysis, and visualization tools to aid the climate research community in searching, exporting, and using datasets.

US CLIVAR contributes to the data-center holdings by promoting the timely migration of quality-controlled, process-study, model-experiment, and field-campaign datasets with appropriate metadata to the centers to facilitate broad community use as a best practices principle.

Ships and aircraft

Regular and high-quality *in situ* measurements are indispensable for observing, monitoring. and understanding the variability of the oceans and overlying atmosphere. This need to collect measurements in and over the oceans makes US CLIVAR research heavily reliant on specialized research vessels and aircraft. The field campaigns and enhanced monitoring efforts discussed earlier would be impossible without these research platforms. US CLIVAR research depends on these assets, and also plays an active role in informing ship and aircraft requirements for climate research.

Within the US, the activities and schedules of federally- owned research vessels, and some privately owned vessels, are coordinated by the University-National Oceanographic Laboratory System (UNOLS), funded by ONR, NSF, NOAA, the US Coast Guard, USGS, and the Minerals Management Service. UNOLS coordinates and reviews the access to and utilization of facilities for oceanographic research and the current match of facilities to the needs of oceanographic programs. It makes recommendations of priorities for replacing, modifying, or improving the numbers and mix of facilities for the community of users. In addition to the 20 research vessels in the UNOLS Fleet, NOAA also operates a fleet of research and operational vessels that occupy repeat hydrographic lines, help maintain a number of components of the sustained Ocean Observing System, and support process studies. The US fleet and research vessels of other nations are absolutely vital to oceanographic and over-ocean atmospheric research—the process studies and much of the climate monitoring efforts of US CLIVAR researchers depend on access to these research vessels.

Research aircraft, which provide platforms for measurements of the atmosphere and properties of the ocean surface, are also used in support of US CLIVAR research for atmospheric, air-sea interaction, and oceanographic campaigns (e.g., Bane et al., 2004). Aircraft assets of NASA, NOAA, NSF, DoE, and ONR are coordinated by the Scientific Committee for Oceanographic Aircraft Research (SCOAR). Established by UNOLS in 2002, SCOAR coordinates and facilitates use of the federal fleet of research aircraft for oceanographic research and provides recommendations to agencies regarding operations, sensor development, fleet composition, fleet utilization, and data services.

Most US CLIVAR field campaigns have utilized research vessels, and many have employed ships and aircraft. The VOCALS-Rex field campaign, discussed earlier, provides one example of coordinated use of research aircraft and vessels for a US CLIVAR process study focused on improving understanding of atmospheric and oceanic processes that play an important role in the climate system (Wood et al., 2011).

Modeling centers and high-performance computing

There are several institutions in the US that are engaged in climate modeling efforts and are supported by various US funding agencies: The Geophysical Fluid Dynamics Laboratory (GFDL) and the National Center for Environmental Prediction (NCEP) are part of the National Oceanic and Atmospheric Administra-

tion (NOAA); the National Center for Atmospheric Research (NCAR) is supported by the National Science Foundation (NSF) and Department of Energy (DOE); and the National Aeronautics and Space Administration (NASA) maintains climate modeling efforts at the Goddard Space Flight Center (GSFC) and at the Goddard Institute for Space Studies (GISS).

Climate modeling efforts at various institutions rely on collaborations with each other and also with the external community including academia, i.e., a broad base of the science community that also has active participation in US CLIVAR efforts. Indeed, the concept and implementation of Climate Process Teams championed by the US CLIVAR has developed a pathway for linking efforts at the major US climate modeling centers with the external scientific community, the goal of which is to hasten advances in climate modeling. The unique role of the CPTs has been to link efforts at the major modeling centers with the scientists at the very outset of a research project thereby easing the transition of model codes and software at a later stage.

Another pathway US CLIVAR community benefits from the activities at the modeling centers is the availability of data from a diverse array of climate simulations run at the centers and provided to the community. Such datasets enable science and further research in advancing understanding of variability and predictability in the climate system. The CMIP5 effort included climate simulations from all US climate modeling centers listed above.

The current generation of Global climate models is run on supercomputers that are also part of the climate modeling centers. These centers often provide necessary computing resources for the US CLIVAR community to engage in climate modeling efforts and conduct simulations to understand various aspects of climate variability.

Operational and real-time information centers

Another enabling infrastructure that advances goals of US CLIVAR, and in turn is benefited by them, is the centers that deliver real-time climate monitoring and prediction information. NOAA is at forefront of developing and delivering climate information on an operational basis. The Climate Prediction Center (CPC) within NOAA provides operational monitoring of the global climate system and predictions on extended to seasonal and interannual timescales. Efforts of centers with a real-time mandate are also complemented by similar efforts at other US institutions, notably the International Research

Institute for Climate and Society (IRI) seasonal prediction efforts at NASA GSFC/Global Modeling and Assimilation Office (GMAO), and NOAA GFDL.

The relationship between the US CLIVAR science community and operational centers is mutually beneficial. Datasets provided by the operational community advance research efforts of US CLIVAR. A particular example is the NCEP/NCAR reanalysis effort currently maintained by CPC that has provided a gridded dataset of atmospheric variability after 1948 and is continually extended forward in real-time (Kalnay et al. 1996). Operational and real-time centers, as part of their seasonal forecasting efforts, also generate hindcast datasets that have been used to advance understanding of climate variability and predictability. CPC has also actively provided forecast support for ground operations related to observational field campaigns organized by the US CLIVAR community, for example, during NAME in 2004 and DYNAMO in 2012.

Operational centers also provide an anchor for transitioning research advances into routine products thereby facilitating delivery of scientific advances for the benefit of society. The mechanism for such research-to-operations (R2O) transition involves testbeds, notably the Climate Test-Bed (CTB), that provide a framework for linking scientists at the operational centers with the external research community. Test beds also provide a means for the research community to utilize the operational framework and to follow operational practices at the very outset of developmental efforts.

International and national climate-change assessments

The United Nations International Panel on Climate Change (IPCC) is charged to provide comprehensive scientific assessments of current scientific, technical, and socio-economic information about the risk of climate change, its potential consequences, and options for adaptation and mitigation. US CLIVAR research activities contribute directly to the Assessment process. Published research articles, including those resulting from US CLIVAR Climate Model Evaluation Project, inform the Assessment Report chapter on scientific aspects of the climate system and climate change. US CLIVAR Working Groups (e.g., the Decadal Predictability Working Group) develop metrics for analyzing CMIP simulations. US CLIVAR-affiliated scientists serve as lead and contributing authors to the Report, while many others participate in its public review.

In the US, a National Climate Assessment (NCA) Report is to be organized and produced by the USGCRP at least every four years as required by the Global Change Research Act of 1990. The NCA surveys, integrates, and synthesizes scientific understanding across disciplines, regions, and resource sectors, with the aim of highlighting knowledge to inform resource management decisions and policy choices. The report draws upon USGCRP research, including US CLIVAR science, to document the past evolution, evaluate the current state, and project future changes in US climate, providing a resource for understanding and communicating climate change science and impacts for the nation. The US CLIVAR program is exploring opportunities to engage and inform the NCA process on an ongoing basis.

References

Alexander, M. A., I. Bladé, M. Newman, J. R. Lanzante, N.C. Lau, and J. D. Scott, 2002: The atmospheric bridge: The influence of ENSO teleconnections on air-sea interaction over the global oceans. *J. Climate*, **15**, 2205-2231, doi:10.1175/1520-0442(2002)015<2205:TABTIO>2.0.CO;2.

Arakawa, A., 2004: The Cumulus Parameterization Problem: Past, Present, and Future. *J. Climate*, **17**, 2493–2525, doi:10.1175/1520-0442(2004)017<2493:RATCPP>2.0.CO;2.

Arblaster, J.M., and L.V. Alexander, 2012: The impact of the El Niño-Southern Oscillation on maximum temperature extremes. *Geophys. Res. Lett.*, **39**, L20702, doi:10.1029/2012GL053409.

Atlas, R., 1997: Atmospheric observations and experiments to assess their usefulness in data assimilation. *J. Met Soc. Japan.*, **75**, 111-130.

Bane, J.M., R. Bluth, C. Flagg, C.A. Friehe, H. Jonsson, W.K. Melville, M. Prince, and D. Riemer, 2004: UNOLS establishes SCOAR to promote research aircraft facilities for US ocean sciences. *Oceanography*, **17**, 176-185, doi:10.5670/oceanog.2004.14.

Blunden, J., and D. S. Arndt, Eds., 2012: State of the Climate in 2011. Bull. Amer. Meteor. Soc., **93** (7), S1-S264.

Böning, C. W., M. Scheinert, J. Dengg, A. Blastoch, and A. Funk, 2006: Decadal variability of subpolar gyre transport and its reverberation in the North Atlantic overturning. *Geophys. Res. Lett.*, **33**, L21S01, doi:10.1029/2006GL026906.

Böning, C. W., A. Dispert, M. Visbeck, S. R. Rintoul and F. U. Schwarzkopf, 2008: The response of the Antarctic Circumpolar Current to recent climate change. Nat. Geosci., **1**, 864-869, doi:10.1038/ngeo362.

Booth, B. B., N. J. Dunstone, P. R. Halloran, T. Andrews, and N. Bellouin, 2012: Aerosols implicated as a prime driver of twentieth-century North Atlantic climate variability. *Nature*, **484**, 228-232, doi:10.1038/nature10946.

Bourassa, Mark A., and Co-authors, 2013: High-Latitude Ocean and Sea Ice Surface Fluxes: Challenges for Climate Research. *Bull. Amer. Meteor. Soc.*, **94**, 403–423. doi:10.1175/BAMS-D-11-00244.1.

Branstator, G., and H. Teng, 2010: Two Limits of Initial-Value Decadal Predictability in a CGCM. *J. Climate*, **23**, 6292–6311. doi:10.1175/2010JCLI3678.1.

Bureau of Reclamation. 2007: Climate Technical Work Group Report: Review of Science and Methods for Incorporating Climate Change Information into Bureau of Reclamation's Colorado River Basin Planning Studies, eds. L. Brekke, B. Harding, T. Piechota, B. Udall, C. Woodhouse, and D. Yates, prepared as Appendix U in: *Final EIS Colorado River Interim Guidelines for Lower Basin Shortages and Coordinated Operations for Lake Powell and Lake Mead*, Bureau of Reclamation, US Department of the Interior, 118pp, http://www.usbr.gov/lc/region/programs/strategies/FEIS/.

CCSP (U.S. Climate Change Science Program), 2008: Reanalysis of Historical Climate Data for Key Atmospheric Features: Implications for Attribution of Causes of Observed Change. *A Report by the U.S. Climate Change Science Program and the Subcommittee on Global Change Research* [Randall Dole, Martin Hoerling, and Siegfried Schubert (eds.)]. National Oceanic and Atmospheric Administration, National Climatic Data Center, Asheville, NC, 156 pp.

Center for Research on Environmental Decisions (CRED), 2009: The Psychology of Climate Change Communication: A Guide for Scientists, Journalists, Educators, Political Aides, and the Interested Public. New York. http://cred.columbia.edu/guide.

Cohen, J., J. C. Furtado, M. Barlow, A. Alexeev, and J. Cherry 2012a: Arctic warming, increasing fall snow cover and widespread boreal winter cooling. *Environ. Res. Lett.*, **7**, 014007, doi:10.1088/1748-9326/7/1/014007.

Cohen, J., J. C. Furtado, M. Barlow, A. Alexeev, and J. Cherry 2012b: Asymmetric seasonal temperature trends. *Geophys. Res. Lett.*, **39**, L04705, doi:10.1029/2011GL050582.

Collins, M., M. Botzet, A. F. Carril, H. Drange, A. Jouzeau, M. Latif, S. Masina, O. H. Otteraa, H. Pohlmann, A. Sorteberg, R. Sutton, and L. Terray, 2006: Interannual to decadal climate predictability in the North Atlantic: A multimodel-ensemble study. *J. Climate*, **19**, 1195–1202, doi:10.1175/JCLI3654.1.

Collins, M., S. I. An, W. Cai, A. Ganachaud, E. Guilyardi, F.-F. Jin, M. Jochum, M. Lengaigne, S. Power, A. Timmermann, G. Vecchi, and A. Wittenberg, 2010: The impact of global warming on the tropical Pacific Ocean and El Niño. *Nature Geosci.*, **3**, 391-397, doi:10.1038/ngeo868.

Cronin, M. F., S. Legg, and P. Zuidema, 2009: Best practices for process studies. *Bull. Amer. Meteorol. Soc*, **90**, 917-918, doi:10.1175/2009BAMS2622.1.

Davis, R.E., W.S Kessler, R. Lukas, R.A. Weller, D.W. Behringer, D.R. Cayan, D.B. Chelton, C. Eriksen, S. Esbensen, R.A. Fine, I. Fukumori, G. Kiladis, M.C. Gregg, E. Harrison, G.C. Johnson, T. Lee, N.J. Mantua, J.P. McCreary, M.J. McPhaden, J.C. McWilliams, A.J. Miller, H. Mitsudera, P.P. Niiler, B. Qiu, D. Raymond, D. Roemmich, D.L. Rudnick, N. Schneider, P.S. Schopf, D. Stammer, L. Thompson, and W.B. White, 2000: *Implementing the Pacific Basin Extended Climate Study (PBECS)*. Washington, D.C.: US CLIVAR Office, 123pp, http://www.usclivar.org/sites/default/files/pbecs-2000.pdf.

Deser, C., M. A. Alexander, S.-P. Xie, and A. S. Phillips, 2010: Sea surface temperature variability: Patterns and mechanisms. *Ann. Rev. Marine Sci.*, **2**, 115-143, doi:10.1146/annurev-marine-120408-151453.

Deser, C., and Co-authors, 2012a: ENSO and Pacific decadal variability in the community climate system model version 4. *J. Climate*, **25**, 2622-2651, doi:10.1175/JCLI-D-11-00301.1.

Deser, C., A. S. Phillips, V. Bourdette, and H. Teng, 2012b: Uncertainty in climate change projections: The role of internal variability. *Climate Dyn.*, **38**, 527-546, doi:10.1007/s00382-010-0977-x.

Di Lorenzo E., N. Schneider, K. M. Cobb, K. Chhak, P. J. S. Franks, A. J. Miller, J. C. McWilliams, S. J. Bograd, H. Arango, E. Curchister, T. M. Powell, and P. Rivere, 2008: North Pacific Gyre Oscillation links ocean climate and ecosystem change. *Geophys. Res. Lett.*, **35**, L08607, doi:10.1029/2007GL032838.

Di Lorenzo, E., J. Fiechter, N. Schneider, A. Bracco, A. J. Miller, P. J. S. Franks, S. J. Bograd, A. M. Moore, A. C. Thomas, W. Crawford, A. Pena and A. J. Hermann, 2009: Nutrient and salinity decadal variations in the central and eastern North Pacific. *Geophysical Research Letters*, **36**, doi:10.1029/2009gl038261.

Di Lorenzo, E., K. M. Cobb, J. C. Furtado, N. Schneider, B. T. Anderson, A. Bracco, M. A. Alexander, and D. J. Vimont, 2010: Central Pacific El Niño and decadal climate change in the North Pacific. *Nature Geosci.*, **3**, 762–765, doi:10.1038/ngeo984.

Doney, S.C., K. Lindsay, K. Caldeira, J.-M. Campin, H. Drange, J.-C. Dutay, M. Follows, Y. Gao, A. Gnanadesikan, N. Gruber, A. Ishida, F. Joos, G. Madec, E. Maier-Reimer, J.C. Marshall, R.J. Matear , P. Monfray, A. Mouchet, R. Najjar, J.C. Orr, G.-K. Plattner, J. Sarmiento, R. Schlitzer, R. Slater, I.J. Totterdell, M.-F. Weirig, Y. Yamanaka, and A. Yool, 2004: Evaluating global ocean carbon models: the importance of realistic physics. *Global Biogeochem. Cycles*, **18**, GB3017, doi:10.1029/2003GB002150.

Dow, K. and G. Carbone, 2007: Climate Science and Decision Making. *Geogr. Compass*, **1**, 302-324, doi:10.1111/j.1749-8198.2007.00036.x.

Easterling, D. R., and M. F. Wehner, 2009: Is the climate warming or cooling? *Geophys. Res. Lett.*, **36**, L08706, doi:10.1029/2009GL037810.

Eden, C., and J. Willebrand, 2001: Mechanism of interannual to decadal variability of the North Atlantic circulation. *J. Climate*, **14**, 2266-2280, doi:10.1175/1520-0442(2001)014<2266:MOITDV>2.0.CO;2.

Enfield, D. B., A. M. Mestas-Nunez, and P. J. Trimble, 2001: The Atlantic Multidecadal Oscillation and its relationship to rainfall and river flows in the continental US *Geophys. Res. Lett.*, **28**, 2077–2080, doi:10.1029/2000GL012745.

Esbensen, S., C. Gudmundson, and T. Mitchel, 2002: *US CLIVAR Pan American Research: A Scientific Prospectus and Implementation Plan*, Washington, D.C.: US CLIVAR Office, 58pp, http://www.usclivar.org/sites/default/files/PanAm_Plan_2002.pdf.

Fischer, A.S., J. Hall, D.E. Harrison, D. Stammer, and J. Benveniste, 2010: "Conference Summary-Ocean Information for Society: Sustaining the Benefits, Realizing the Potential" in *Proceedings of OceanObs'09: Sustained Ocean Observations and Information for Society (Vol. 1)*, Venice, Italy, 21-25 September 2009, Hall, J., Harrison, D.E. & Stammer, D., Eds., ESA Publication WPP-306, doi:10.5270/OceanObs09.Summary

Francis, J. A., and S. J. Vavrus, 2012: Evidence linking Arctic amplification to extreme weather in mid-latitudes. *Geophys. Res. Lett.*, **39**, L06801, doi:10.1029/2012GL051000.

Frankignoul, C., 1985: Sea surface temperature anomalies, planetary waves, and air-sea feedback in the middle latitudes. *Rev. Geophys.*, **23**, 357-390, doi:10.1029/RG023i004p00357.Friedlingstein, P., P. Cox, R. Betts, L. Bopp, W. Von Bloh, and Co-authors, 2006: Climate-carbon cycle feedback analysis: results from the C4MIP model intercomparison. *J. Climate*, **19**, 3337–3353, doi:10.1175/JCLI3800.1.

Fu, X. H., J.-Y. Lee, P.-C. Hsu, H. Taniguchi, B. Wang, W. Q. Wang, and S. Weaver, 2013: Multi-model MJO forecasting during DYNAMO/CINDY period. *Climate Dyn.*, **41**, 1067-1081, doi:10.1007/s00382-013-1859-9.

Furtado, J. C., E. Di Lorenzo, N. Schneider, and N. A. Bond, 2011: North Pacific decadal variability and climate change in the IPCC AR4 models. *J. Climate*, **24**, 3049-3067, doi:10.1175/2010JCLI3584.1.

Graham, N. E., 1994: Decadal-scale climate variability in the tropical and North Pacific during the 1970s and 1980s: Observations and model results. *Climate Dyn.*, **10**, 135–162, doi:10.1007/BF00210626.

Grotjahn, R., 2013: Ability of CCSM4 to simulate California extreme heat conditions from evaluating simulations of the associated large scale upper air pattern. *Climate Dyn.*, **41**, 1187-1197, doi:10.1007/s00382-013-1668-1.

Gruber, N., M. Gloor, S. E. Mikaloff Fletcher, S. Doney, S. Dutkiewicz, M. J. Follows, M. Gerber, A. R. Jacobson, F. Joos, K. Lindsay, D. Menemenlis, A. Mouchet, S. A. Muller, J. L. Sarmiento, and T. Takahashi, 2009: Oceanic sources, sinks, and transport of atmospheric CO_2, *Global Biogeochem. Cycles*, **23**, GB1005, doi:10.1029/2008GB003349.

Gu, D., and S.G.H. Philander, 1997: Interdecadal climate fluctuations that depend on exchanges between the tropics and the extratropics. *Science*, **275**, 805-807, doi:10.1126/science.275.5301.805.

Guilyardi E., W. Cai, M. Collins, A. Fedorov, F.-F. Jin, A. Kumar, D.-Z. Sun, and A. Wittenberg, 2012: CLIVAR workshop summary: New strategies for evaluating ENSO processes in climate models. *Bull. Amer. Met. Soc.*, **93**, 235-238, doi:10.1175/BAMS-D-11-00106.1.

Hallberg, R.W., and A. Gnanadesikan, 2006: The role of eddies in determining the structure and response of the wind-driven Southern Hemisphere overturning: Initial results from the Modeling Eddies in the Southern Ocean (MESO) Project, *J. Phys. Oceanogr.*, **36**, 2232-2252, doi:10.1175/JPO2980.1.

Hawkins, E., and R. Sutton, 2009: The potential to narrow uncertainty in regional climate predictions. *Bull. Amer. Meteor. Soc.*, **90**, 1095–1107, doi:10.1175/2009BAMS2607.1.

Hawkins, E., J. Robson, R. Sutton, D. Smith, and N. Keenlyside, 2011: Evaluating the potential for statistical decadal predictions of sea surface temperatures with a perfect model approach. *Climate Dyn.*, **37**, 2495–2509, doi:10.1007/s00382-011-1023-3.

Held, I. M., 2005: The Gap between simulation and understanding in climate modeling. *Bull. Amer. Meteor. Soc.*, **86**, 1609–1614, doi:10.1175/BAMS-86-11-1609.

Higgins, R. W., J. E. Schemm, W. Shi, and A. Leetmaa, 2000: Extreme precipitation events in the western United States related to tropical forcing. *J. Climate*, **13**, 793-820, doi:10.1175/1520-0442(2000)013<0793:EPEITW>2.0.CO;2.

Higgins, W., D., and Co-authors, 2006: The NAME 2004 field campaign and modeling strategy. *Bull. Amer. Meteor. Soc.*, **87**, 79–94, doi:10.1175/BAMS-87-1-79.

Hoerling, M. P., A. Kumar, J. Eischeid, and B. Jha, 2008: What is causing the variability in global land temperature? *Geophys. Res. Lett.*, **35**, L23712, doi:10.1029/2008GL035984.

Hoose, C., J. E. Kristjánsson, T. Iversen, A. Kirkevåg, Ø. Seland, and A. Gettelman, 2009: Constraining cloud droplet number concentration in GCMs suppresses the aerosol indirect effect. *Geophys. Res. Lett.*, **36**, L12807, doi:10.1029/2009GL038568.

Houghton, J., J. Townshend, K. Dawson, P. Mason, J. Zillman, and A. Simmons, 2012: GCOS at 20 years: the origin, achievement and future development of the Global Climate Observing System. *Weather*, **67**, 227-235, doi: 10.1002/wea.1964.

Hurrell, J. W., 1995: Decadal trends in the North Atlantic Oscillation: regional temperatures and precipitation. *Science*, **269**, 676-679, doi:10.1126/science.269.5224.676.

Hwang, Y.-T., and D. M. W. Frierson, 2013: Link between the double-Intertropical Convergence Zone problem and cloud bias over Southern Ocean. *Proc. Nat. Acad. Sci.*, **110**, 4935–4940, doi:10.1073/pnas.1213302110.

Imada, Y., and M. Kimoto, 2012: Parameterization of Tropical Instability Waves and Examination of Their Impact on ENSO Characteristics. *J. Climate*, **25**, 4568–4581, doi:10.1175/JCLI-D-11-00233.1.

Johannessen, O. M., L. Bengtsson, M. W. Miles, S. I. Kuzmina, V. A. Semenov, G. V. Alekseev, A. P. Nagurnyi, V. F. Zakharov, L. P. Bobylev, L. H. Pettersson, K. Hasselmann, and H. P. Cattle, 2004: Arctic climate change: observed and modelled temperature and sea-ice variability. *Tellus A*, **56**, 328–341, doi: 10.1111/j.1600-0870.2004.00060.x

Joyce, T., and J. Marshall, 2000: *Implementation Plan for Atlantic Climate Variability Experiment: Summary and Recommendations.* Washington, D.C.: US CLIVAR Office, 21pp, www.usclivar.org/sites/default/files/ACVE_Implementation_draft.pdf.

Kalnay, E., M. Kanamitsu, R. Kistler, W. Collins, D. Deaven, L. Gandin, M. Iredell, S. Saha, G. White, J. Woollen, Y. Zhu, M. Chelliah, W. Ebisuzaki, W.Higgins, J. Janowiak, K. C. Mo, C. Ropelewski, J. Wang, A. Leetmaa, R. Reynolds, R. Jenne, and D. Joseph, 1996: The NCEP/NCAR 40-year reanalysis project. *Bull. Amer. Meteor. Soc*, **77**, 437–471, doi:10.1175/1520-0477(1996)077<0437:TNYRP>2.0.CO;2.

Kamenkovich, I., and T. Radko, 2011: Role of the Southern Ocean in setting the Atlantic stratification and meridional overturning circulation. *J. Marine Res.*, **69**, 277-308, doi:10.1357/002224011798765286.

Karentz, D., and I. Bosch, 2001: Influence of ozone-related increases of ultraviolet radiation on Antarctic marine organisms. *Am. Zoo.*, **41**, 3-16, doi:10.1093/icb/41.1.3.

Kaufmann, R. K., H. Kauppi, M. L. Mann, and J. H. Stock, 2011: Reconciling anthropogenic climate change with observed temperature 1998–2008. *Proc. Nat. Acad. Sci.*, **108**, 11790–11793,doi:10.1073/pnas.1102467108.

Kessler, W. S., and R. Kleeman, 2000: Rectification of the Madden-Julian Oscillation into the ENSO cycle. *J. Climate*, **13**, 3560-3575, doi:10.1175/1520-0442(2000)013<3560:ROTMJO>2.0.CO;2.

Kiehl, J. T., 2007: Twentieth century climate model response and climate sensitivity. *Geophys. Res. Lett.*, **34**, L22710, doi:10.1029/2007GL031383.

Kirtman, B., C. M. Bitz, F. Bryan, W. Collins, J. Dennis, N. Hearn, J. L. Kinter III, R. Loft, C. Rousset, L. Siqueira, Cr. Stan, R. Tomas, and M. Vertenstein, 2012: Impact of Ocean Model Resolution on CCSM Climate Simulations. *Climate Dyn.*, **39**, 1303-1328, doi:10.1007/s00382-012-1500-3.

Kleeman, R., J. P. McCreary, and B. A. Klinger 1999: A mechanism for generating ENSO decadal variability. *Geophys. Res. Lett.*, **26**, 1743-1746, doi:10.1029/1999GL900352.

Klein, Stephen A., Y. Zhang, M. D. Zelinka, R. N. Pincus, J. Boyle, and P. J. Gleckler, 2013: Are climate model simulations of clouds improving? An evaluation using the ISCCP simulator. *J. Geophys. Res.*, **118**, 1329–1342, doi:10.1002/jgrd.50141.

Knight, J. R., R. J. Allan, C. K. Folland, M. Vellinga, and M. E. Mann, 2005: A signature of persistent natural thermohaline circulation cycles in observed climate. *Geophys. Res. Lett.*, **32**, L20708, doi:10.1029/2005GL024233.

Kumar, A., B. Jha, Q. Zhang, and L. Bounoua, 2007: A new methodology for estimating the unpredictable component of seasonal atmospheric variability. *J. Climate*, **20**, 3888-3901, doi:10.1175/JCLI4216.1.

Kumar, A., M. Chen, L. Zhang, W. Wang, Y. Xue, C. Wen, L. Marx, and B. Huang, 2012: An analysis of the nonstationarity in the bias of sea surface temperature forecasts for the NCEP Climate Forecast System (CFS) Version 2. *Mon. Wea. Rev.*, **140**, 3003–3016, doi:10.1175/MWR-D-11-00335.1.

Lau, W. K., and D. Waliser, 2012: *Intraseasonal variability in the atmosphere-ocean climate system*. Springer, 613pp.

Lee, T., 2004: Decadal weakening of the shallow overturning circulation in the South Indian Ocean. *Geophys. Res. Lett.*, **31**, L18305, doi:10.1029/2004GL020884.

Lee, T., and M.J. McPhaden 2008: Decadal phase change in large-scale sea level and winds in the Indo-Pacific region at the end of the 20th century. *Geophys. Res. Lett.*, **35**, L01605, doi:10.1029/2007GL032419.

Lemke, P., Ren, J., Alley, R., Allison, I., Carrasco, J., Flato, G., Fujii, Y., Kaser, G., Mote, P., Thomas, R. and Zhang, T. (2007). Chapter 4: Observations: Changes in Snow, Ice and Frozen Ground. In Climate Change 2007: The Physical Science Basis. Contribution of Working Group 1 to the Fourth Assessment Report of the Intergovernmental Panel on Climate Change (eds. S. Solomon, D. Qin, M. Manning, Z. Chen, M.C. Marquis, K. Averyt, M. Tignor and H.L. Miller). Intergovernmental Panel on Climate Change, Cambridge and New York.

Lemus-Deschamps L., and J. Makin, 2012: Fifty years of changes in UV index and implications for melanoma skin cancer in Australia. *Int. J. Biometeor.*, **56**, 727-735, doi:10.1007/s00484-011-0474-x.

Lindsay, R. W., J. Zhang, A. J. Schweiger, M. A. Steele, and H. Stern, 2009: Arctic sea ice retreat in 2007 follows thinning trend. *J. Climate*, **22**, 165-176, doi:10.1175/2008JCLI2521.

Lindstrom, E., J. Gunn, A. Fischer, A. McCurdy, L. K. Glover, and the Task Team for an Integrated Framework for Sustained Ocean Observing, 2012: A Framework for Ocean Observing. UNESCO 2012, IOC/INF-1284, doi:10.5270/OceanObs09-FOO.

Liu, J., B. Wang, M. Cane, S.-Y. Kim, and J.-Y., Lee, 2013: Divergent global precipitation changes induced by natural versus anthropogenic forcing. *Nature*, **493**, 656–659, doi:10.1038/nature11784.

Liu, Z., 2012: Dynamics of interdecadal climate variability: A historical perspective. *J. Climate*, **25**, 1963-1995, doi:10.1175/2011JCLI3980.1.

Lovenduski, N. S., N. Gruber, and S. C. Doney, 2008: Toward a mechanistic understanding of the decadal trends in the Southern Ocean carbon sink. *Global Biogeochem. Cycles*, **22**, GB3016, doi:10.1029/2007GB003139.

Lovenduski, N. S., and T. Ito, 2009: The future evolution of the Southern Ocean CO_2 sink. *J. Mar. Res.*, **67**, 597-617, doi:10.1357/002224009791218832.

Lowrey, J. L., A. J. Ray, and Co-authors, 2009: Factors influencing the use of climate information by Colorado municipal water managers. *Climate Res.*, **40**, 103–119, doi:10.3354/cr00827.

Mahowald, N., D. Ward, S. Kloster, M. Flanner, C. Heald, N. Heavens, P. Hess, J.-F. Lamarque, and P. Chuang, 2011: Aerosol impacts on climate and biogeochemistry, *Ann. Rev. Environ. Resour.*, **36**, 45-74, doi:10.1146/annurev-environ-042009-094507.

Maloney, E. D., and D. L. Hartmann, 2000: Modulation of eastern North Pacific hurricanes by the Madden-Julian oscillation. *J. Climate*, **13**, 1451-1460, doi:10.1175/1520-0442(2000)013<1451:MOENPH>2.0.CO;2.

Mantua, N. J., S. R. Hare, Y. Zhang, J. M. Wallace, R. C. Francis, 1997: A Pacific interdecadal climate oscillation with impacts on salmon production. *Bull. Amer. Meteor. Soc.*, **78**, 1069–1079, doi:10.1175/1520-0477(1997)078<1069:APICOW>2.0.CO;2.

Mantua, N., I. Tohver, and A. Hamlet, 2010: Climate change impacts on streamflow extremes and summertime stream temperature and their possible consequences for freshwater salmon habitat in Washington State. *Climate Change*, **102**, 187-223, doi:10.1007/s10584-010-9845-2.

Marshall, J., and T. Radko, 2003: Residual-mean solutions for the Antarctic Circumpolar Current and its associated overturning circulation. *J. Phys. Oceanogr.*, **33**, 2341-2354, doi:10.1175/1520-0485(2003)033<2341:RSFTAC>2.0.CO;2.

Matsumoto, K., J. L. Sarmiento, R. M. Key, J. L. Bullister, K. Caldeira, J.-M. Campin, S. C. Doney, H. Drange, J.-C. Dutay, M. Follows, Y. Gao, A. Gnanadesikan, N. Gruber, A. Ishida, F. Joos, K. Lindsay, E. Maier-Reimer, J. C. Marshall, R. J. Matear, P. Monfray, R. Najjar, G.-K. Plattner, R. Schlitzer, R. Slater, P. S. Swathi, I. J. Totterdell, M.-F. Weirig, Y. Yamanaka, A. Yool, and J. C. Orr, 2004: Evaluation of ocean carbon cycle models with data-based metrics. *Geophys. Res. Lett.*, **31**, L07303, doi:10.1029/2003GL018970.

McCarthy, J. J., O. F. Canziani, N. A. Leary, D. J. Dokken, and K. S. White (Eds.), 2001: *Climate change 2001: impacts, adaptation, and vulnerability: contribution of Working Group II to the third assessment report of the Intergovernmental Panel on Climate Change.* Cambridge University Press.

McFadden, E. M., I. M. Howat, I. Joughin, B. Smith, and Y. Ahn, 2011: Changes in the dynamics of marine terminating outlet glaciers in west Greenland (2000-2009). *J. Geophys. Res.*, **116**, F02022, doi:10.1029/2010JF001757.

McKeon, G. M., G. S. Stone, J. I. Syktus, J. O. Carter, N. R. Flood, D. G. Ahrens, D. N. Bruget, C. R. Chilcott, D. H. Cobon, R. A. Cowley, S. J. Crimp, G. W. Fraser, S. M. Howden, P. W. Johnston, J. G. Ryan, C. J. Stokes, and K. A. Day, 2009: Climate change impacts on northern Australian rangeland livestock carrying capacity: a review of issues. *Rangeland Journal*, **31**, 1-29, doi:10.1071/RJ08068.

McPhaden, M. J., and Co-authors, 2010: A TOGA Retrospective. *Oceanography*, **23**, 86-103, doi:10.5670/oceanog.2010.26.

Meehl, G., and C. Tebaldi, 2004: More intense, more frequent and longer lasting heat waves in the 21st century. *Science*, **305**, 994-997, doi:10.1126/science.1098704.

Meehl, G. A., J. M. Arblaster, J. T. Fasullo, A. Hu, and K. E. Trenberth, 2011: Model-based evidence of deep-ocean heat uptake during surface-temperature hiatus periods. *Nat. Climate Change*, doi:10.1038/nclimate1229.

Mehta, V., and Co-authors, 2011: Decadal Climate Predictability and Prediction: Where Are We?. *Bull. Amer. Meteor. Soc.*, **92**, 637–640, doi:10.1175/2010BAMS3025.1.

Meinke, H., P. deVoil, G. L. Hammer, S. Power, R. Allan, R. C. Stone, C. Folland, A. Potgieter, 2005: Rainfall Variability at Decadal and Longer Time Scales: Signal or Noise? *J. Climate*, **18**, 89-96, doi:10.1175/JCLI-3263.1.

Meredith, M. P., and Co-authors, 2011: Sustained monitoring of the Southern Ocean at Drake Passage: Past achievements and future priorities. *Rev. Geophys.*, **49**, RG4005, doi:10.1029/2010RG000348.

Moon T., I. Joughin, B. Smith, and I. Howat, 2012: 21st-Century Evolution of Greenland Outlet Glacier Velocities. *Science*, **336**, 576-578, doi:10.1126/science.1219985.

Murphy, J. M., D. M. Sexton, D. N. Barnett, G. S. Jones, M. J. Webb, M. Collins, and D. A. Stainforth, 2004: Quantification of modelling uncertainties in a large ensemble of climate change simulations. *Nature*, **430**, 768-772, doi:10.1038/nature02771.

National Research Council (NRC), 1994: *GOALS (Global Ocean-Atmosphere-Land System) for Predicting Seasonal-to-Interannual Climate: A Program of Observation, Modeling, and Analysis.* Washington, D.C.: National Academy Press, 116pp.

National Research Council (NRC), 1995: *Natural Climate Variability on Decadal to Century Time Scales.* D. G. Martinson, K. Bryan, M. Ghil, M. M. Hall, T. R. Karl, E. S. Sarachik, S. Sorooshian, and L. D. Talley (eds.). Washington, D.C.: National Academy Press, 644 pp.

National Research Council (NRC), 1998a: *A Scientific Strategy for US Participation in the GOALS (Global Ocean-Atmosphere-Land System) Component of the CLIVAR (Climate Variability and Predictability) Programme.* Washington, D.C.: National Academy Press, 88pp.

National Research Council (NRC), 1998b: *Decade-to-Century-Scale Climate Variability and Change: A Science Strategy.* Washington, D.C.: National Academy Press, 160pp.

National Climate Data Center (NCDC), 2012: *Billion-Dollar Weather/Climate Events*, Available online: http://www.ncdc.noaa.gov/billions.

Newman, M., 2007: Interannual to decadal predictability of tropical and North Pacific sea surface temperatures. *J. Climate*, **20**, 2333–2356, doi:10.1175/JCLI4165.1.

Newman, M., G. P. Compo, and M. A. Alexander, 2003: ENSO-forced variability of the Pacific Decadal Oscillation. *J. Climate*, **16**, 3853-3857, doi:10.1175/1520-0442(2003)016<3853:EVOTPD>2.0.CO;2.

Nonaka, M., S.-P. Xie, and J. P. McCreary, 2002: Decadal variations in the Subtropical Cells and equatorial Pacific SST. *Geophys. Res. Lett.*, **29**, doi:10.1029/2001GL013676.

Orr, J. C., E. Maier-Reimer, U. Mikolajewicz, P. Monfray, J. L. Sarmiento, J. R. Toggweiler, N. K. Taylor, J. Palmer, N. Gruber, C. L. Sabine, C. Le Quéré, R. M. Key, and J. Boutin, 2001: Estimates of anthropogenic carbon uptake from four three-dimensional global ocean models. *Global Biogeochem. Cycles*, **15**, 43-60, doi:10.1029/2000GB001273.

Palmer, T. N., F. J. Doblas-Reyes, A. Weisheimer, M. J. Rodwell, 2008: Toward Seamless Prediction: Calibration of Climate Change Projections Using Seasonal Forecasts. *Bull. Amer. Meteor. Soc.*, **89**, 459–470, doi:10.1175/BAMS-89-4-459.

Pegion, K. and P. D. Sardeshmukh, 2011: Prospects for Improving Subseasonal Predictions. *Mon. Wea. Rev.*, **139**, 3648–3666. doi: http://dx.doi.org/10.1175/MWR-D-11-00004.1.

Power, S., T. Casey, C. Folland, A. Colman, and V. Mehta, 1999a: Interdecadal modulation of the impact of ENSO on Australia, *Climate Dyn.*, **15**, 319–324, doi:10.1007/s003820050284.

Power, S., F. Tseitkin, V. Mehta, S. Torok, and B. Lavery 1999b: Decadal climate variability in Australia during the 20th century. *Int. J. Climatol.*, **19**, 169-184, doi:10.1002/(SICI)1097-0088(199902)19:2<169::AID-JOC356>3.0.CO;2-Y.

Riebesell, U., Schulz, K. G., Bellerby, R. G. J. et al., 2007: Enhanced biological carbon consumption in high CO_2 ocean. *Nature*, **450**, 545–548, doi:10.1038/nature06267.

Rignot, E., I. Velicogna, M. R. van den Broeke, A. Monaghan, and J. T. M. Lenaerts, 2011: Acceleration of the contribution of the Greenland and Antarctic ice sheets to sea level rise. *Geophys. Res. Lett.*, **38**, L05503, doi:10.1029/2011GL046583.

Rintoul, S., and Co-authors, 2010: "Deep Circulation and Meridional Overturning: Recent Progress and a Strategy for Sustained Observations" in Proceedings of OceanObs'09: Sustained Ocean Observations and Information for Society (Vol. 1), Venice, Italy, 21-25 September 2009, Hall, J., Harrison, D.E. & Stammer, D., Eds., ESA Publication WPP-306, pp.32, doi:10.5270/OceanObs09.

Robson, J., R. Sutton, K. Lohmann, D. Smith, and M. D. Palmer, 2012: Causes of the rapid warming of the North Atlantic Ocean in the mid-1990s. *J Clim*ate, **25**, 4116–4134, doi:10.1175/JCLI-D-11-00443.1.

Rowley, R. J., J. C. Kostlenick, D. Braaten, X. Li, and J. Meisel, 2007: Risk of rising sea level to population and land area. *EOS Transactions*, **88**, 105-107, doi:10.1029/2007EO090001.

Russell, J. L., R. J. Stouffer, and K. W. Dixon, 2006: Intercomparison of the Southern Ocean circulations in the IPCC coupled model control simulations. *J. Climate*, **19**, 4560-4575, doi:10.1175/JCLI3869.1.

Sabine, C. L., and Co-authors, 2004: The ocean sink for anthropogenic CO_2. *Science*. **305**. 367-371, doi:10.1126/science.1097403.

Sarmiento, J. L., T. M. C. Hughes, R. J. Stouffer, and S. Manabe, 1998: Simulated response of the ocean carbon cycle to anthropogenic climate warming. *Nature*, **393**, 245-249, doi:10.1038/30455.

Sarmiento, J. L., N. Gruber, M. Brzezinksi, and J. Dunne, 2004: High latitude controls of thermocline nutrients and low latitude biological productivity. *Nature*, **426**, 56-60, doi:10.1038/nature02127.

Scaife, A., D. Copsey, C. Gordon, C. Harris, T. Hinton, S. Keeley, A. O'Neil, M. Roberts, and K. Williams, 2011: Improved Atlantic winter blocking in a climate model. *Geophys. Res. Lett.*, **38**, L23703, doi:10.1029/2011GL049573.

Schott, F. A., J. P. McCreary, and G. C. Johnson, 2004: Shallow overturning circulations of the tropical subtropical oceans. *Earth's Climate: The Ocean–Atmosphere Interaction, Geophys. Monogr.*, No. 147, AGU, 1-16, doi:10.1029/147GM15.

Schweiger, A., R. Lindsay, J. Zhang, M. Steele, H. Stern, and R. Kwok, 2011: Uncertainty in Modeled Arctic Sea Ice Volume. *J. Geophys. Res.*, **116**, C00D06, doi:10.1029/2011JC007084.

Smith, A., and R. Katz, 2013: US Billion-dollar Weather and Climate Disasters: Data Sources, Trends, Accuracy and Biases. *Nat. Hazards*, **67**, 387-410, doi:10.1007/s11069-013-0566-5.

Smith, D. M., and Co-authors, 2013: Real-time multi-model decadal climate predictions. *Climate Dyn.*, online first, doi:10.1007/s00382-012-1600-0.

Smith, K. L., Jr., H. A. Ruhlb, B. J. Bettb, D. S. M. Billettb, R. S. Lampittb, and R. S. Kaufmann, 2009: Climate, carbon cycling, and deep-ocean ecosystems. *Proc. Nat. Acad. Sci.*, **106**, 19211-19218, doi:10.1073/pnas.0908322106.

Solomon, A., J. P. McCreary, R. Kleeman, and B.A. Klinger, 2003: Interannual and decadal variability in an intermediate coupled model of the Pacific region. *J. Climate*, **16**, 383-405, doi:10.1175/1520-0442(2003)016<0383:IADVIA>2.0.CO;2.

Solomon, A., and Co-authors, 2011: Distinguishing the roles of natural and anthropogenically forced decadal climate variability: Implications for prediction. *Bull. Amer. Met. Soc.*, **92**, 141-156, doi:10.1175/2010BAMS2962.1.

SPARC, 2010: Usable Science: a Handbook for Science Policy Decision Makers. Boulder, CO: Science Policy Assessment and Research on Climate (SPARC), 19 pp. http://science-policy.colorado.edu/sparc/outreach/sparc_handbook

Srokosz, M., M. Baringer, H. Bryden, S. Cunningham, T. Delworth, S. Lozier, J. Marotzke, and R. Sutton, 2012: Past, Present, and Future Changes in the Atlantic Meridional Overturning Circulation. *Bull. Amer. Meteor. Soc.*, **93**, 1663–1676, doi:10.1175/BAMS-D-11-00151.1.

Stevenson, S., B. Fox-Kemper, M. Jochum, B. Rajagopalan, and S. G. Yeager, 2010: ENSO model validation using wavelet probability analysis. *J. Climate*, **23**, 5540-5547, doi:10.1175/2010JCLI3609.1.

Stevenson, S., B. Fox-Kemper, M. Jochum, R. Neale, C. Deser, and G. Meehl, 2012: Will there be a significant change to El Niño in the twenty-first century? *J. Climate*, **25**, 2129-2145, doi:10.1175/JCLI-D-11-00252.1.

Subramanian, A., M. Jochum, A. J. Miller, R. Neale, H. Seo, D. Waliser, and R. Murtugudde, 2013: The MJO and global warming: A study in CCSM4. *Climate Dyn*, online first, doi:10.1007/s00382-013-1846-1.

Taylor, K. E., R. J. Stouffer, and G. A. Meehl, 2012: An overview of CMIP5 and the experiment design. *Bull. Amer. Meteor. Soc.*, **93**, 485–498, doi:10.1175/BAMS-D-11-00094.1.

Teng, H., G. Branstator, G. Meehl, 2011: Predictability of the Atlantic Overturning Circulation and Associated Surface Patterns in Two CCSM3 Climate Change Ensemble Experiments. *J. Climate*, **24**, 6054-6075, doi:10.1175/2011JCLI4207.1.

Thompson, D. W. J., S. Solomon, 2002: Interpretation of Recent Southern Hemisphere Climate Change. *Science*, **296**, 895-899, doi:10.1126/science.1069270.

Ting, M., Y. Kushnir, R. Seager, and C. Li, 2011: Robust features of Atlantic multi-decadal variability and its climate impacts. *Geophys. Res. Lett.*, **38**, doi:10.1029/2011GL048712.

Toggweiler, J. R., and B. Samuels, 1998: On the ocean's large-scale circulation near the limit of no vertical mixing. *J. Phys. Oceanogr.*, **28**, 1832–1852, doi:10.1175/152004852.

Trenberth, K. E., G. W. Branstator, D. Karoly, A. Kumar, N.-C. Lau, and C. Ropelewski, 1998: Progress during TOGA in understanding and modeling global teleconnections associated with tropical sea surface temperatures. *J. Geophys. Res.*, **107**, C7, 14291-14324, doi:10.1029/97JC01444.

Trenberth, K. E., and J. T. Fasullo, 2010: Simulation of present day and 21st century energy budgets of the southern oceans. *J. Climate*, **23**, 440-454, doi:10.1175/2009JCLI3152.1.

Trenberth, K. E., J. T. Fasullo, and J. Kiehl, 2009: Earth's global energy budget. *Bull. Amer. Meteor. Soc.*, **90**, 311–323. doi:10.1175/2008BAMS2634.1.

Ummenhofer, C. C., 2009: What causes southeast Australia's worst droughts? *Geophys. Res. Lett.*, **36**, L04706, doi:10.1029//2008GL036801.

US CLIVAR ENSO Diversity Working Group, 2013: Report on the ENSO Diversity Workshop. US CLIVAR Report No. 2011-1, US CLIVAR Project Office, 20pp., http://www.usclivar.org/sites/default/files/meetings/ENSO_Diversity_Workshop_Report.pdf.

US CLIVAR Project Office, 2013: US CLIVAR Accomplishments Report. Report 2013-6, US CLIVAR Project Office, Washington, DC, 20002.

US CLIVAR Scientific Steering Committee (SSC), 2000: *Implementing US CIVAR 2001-2015*. Washington, D.C.: US CLIVAR Office, 75pp.

USGCRP, 2012: *The National Global Change Research Plan 2012-2021: A Strategic Plan for the US Global Change Research Program*, National Coordination Office, 132 pp. http://downloads.globalchange.gov/strategic-plan/2012/usgcrp-strategic-plan-2012.pdf

Vimont, D. J., J. M. Wallace, and D. S. Battisti, 2003: The Seasonal Footprinting Mechanism in the Pacific: Implications for ENSO. *J. Climate*, **16**, 2668-2675, doi:10.1175/1520-0442(2003)016<2668:TSFMIT>2.0.CO;2.

Waliser, D. E., and Co-authors, 2012: The "Year" of Tropical Convection (May 2008–April 2010): Climate Variability and Weather Highlights. *Bull. Amer. Meteor. Soc.*, **93**, 1189–1218, doi:10.1175/2011BAMS3095.1.

Wittenberg, A. T., 2009: Are historical records sufficient to constrain ENSO simulations? *Geophys. Res. Lett.*, **36**, L12702, doi:10.1029/2009GL038710.

Wolfe, C. L., and P. Cessi, 2010: What sets the strength of the middepth stratification and overturning circulation in eddying ocean models? *J. Phys. Oceanogr.*, **40**, 1520-1538, doi:10.1175/2010JPO4393.1.

Wood, R., C. R. Mechoso, C. S. Bretherton, R. A. Weller, B. Huebert, F. Straneo, B. A. Albrecht, H. Coe, G. Allen, G. Vaughan, P. Daum, C. Fairall, D. Chand, L. Gallardo Klenner, R. Garreaud, C. Grados, D. S. Covert, T. S. Bates, R. Krejci, L. M. Russell, S. de Szoeke, A. Brewer, S. E. Yuter, S. R. Springston, A. Chaigneau, T. Toniazzo, P. Minnis, R. Palikonda, S. J. Abel, W. O. J. Brown, S. Williams, J. Fochesatto, J. Brioude, and K. N. Bower, 2011: The VAMOS Ocean-Cloud-Atmosphere-Land Study Regional Experiment (VOCALS-REx): goals, platforms, and field operations. *Atmos. Chem. Phys.*, **11**, 627-654, doi:10.5194/acp-11-627-2011.

World Climate Research Program (WCRP), 1995: *CLIVAR A Study on Climate Variability and Predictability Science Plan*. WCRP No. 89. WMO/TD No. 690. Geneva: WCRP, 172 pp.

WCRP, 2010: *CLIVAR Scientific Frontiers and Imperatives*, World Climate Research Program, 41pp. http://www.clivar.org/sites/default/files/Frontiers%26Imperatives.pdf

WCRP, 2012: *The Climate Variability and Predictability (CLIVAR) Handbook*, World Climate Research Program, 23 pp. http://www.clivar.org/sites/default/files/Handbook_August_12_0.pdf

Wunsch, C., and R. Ferrari, 2004: Vertical mixing, energy, and the general circulation of the ocean. *Ann. Rev. Fluid Mech.*, **36**, 281–314, doi:10.1146/annurev.fluid.36.050802.122121.

Yu, J.-Y., and C. R. Mechoso, 1999: Links between annual variations of Peruvian stratocumulus clouds and of SST in the eastern equatorial Pacific. *J. Climate*, **12**, 3305–3318, doi:10.1175/1520-0442(1999)012<3305:LBAVOP>2.0.CO;2.

Zhang, C., 2005: Madden-Julian Oscillation. *Rev. Geophys.*, **43**, RG2003, doi:10.1029/2004RG000158.

Zhang, C., J. Gottschalck, E. D. Maloney, M. W. Moncrieff, F. Vitart, D. E. Waliser, B. Wang, and M. C. Wheeler, 2013: Cracking the MJO Nut. *Geophys. Res. Lett.*, **40**, doi:10.1002/grl.50244.

Zhang, J., 2007: Increasing Antarctic sea ice under warming atmospheric and oceanic conditions. *J. Climate*, **20**, 2515–2529, doi:10.1175/JCLI4136.1.

Zhang, R., 2008: Coherent surface-subsurface fingerprint of the Atlantic meridional overturning circulation. *Geophys. Res. Lett.*, **35**, doi:10.1029/2008GL035463.

Zhang, R., and T. L. Delworth, 2006: Impact of Atlantic multidecadal oscillations on India/Sahel rainfall and Atlantic hurricanes. *Geophys. Res. Lett.*, **33**, doi:10.1029/2006GL026267.

Zhang, R., T. L. Delworth, R. Sutton, D. Hodson, K. W Dixon, I. M Held, Y. Kushnir, J. Marshall, Y. Ming, R. Msadek, J. Robson, A. Rosati, M. Ting, and G. A. Vecch, 2013: Have aerosols caused the observed Atlantic multidecadal variability? J. Atmos. Sci., **70**, 1135–1144. doi:http://dx.doi.org/10.1175/JAS-D-12-0331.1.

Zhou, L., and R. Murtugudde, 2009: Ocean-atmosphere coupling on different temporal and spatial scales. *J. Atmos. Sci.*, **66**, 1834-1844, doi:10.1175/2008JAS2879.1.

Zhou, S., and A. J. Miller, 2005: The interaction of the Madden-Julian Oscillation and the Arctic Oscillation. *J. Climate*, **18**, 143-159, doi:10.1175/JCLI3251.1.

Appendix A:
Contributors

Science Plan Lead Authors:

Lisa Goddard	Columbia University/International Research Institute for Climate and Society
Baylor Fox-Kemper	Brown University
Arun Kumar	NOAA National Centers for Environmental Prediction
Jay McCreary	University of Hawaii
Michael Patterson	US CLIVAR Project Office
Janet Sprintall	University of California, San Diego/Scripps Institution of Oceanography
Robert Wood	University of Washington

Members of the US CLIVAR SSC, acknowledged below, have overseen the development of the Science Plan, including its drafting, review, and editing:

Bruce Anderson	Boston University
Nicholas Bond	University of Washington/NOAA Pacific Marine Environmental Laboratory
Michael Bosilovich	NASA Goddard Space Flight Center
Annalisa Bracco	Georgia Tech
J. Tom Farrar	Woods Hole Oceanographic Institution
Baylor Fox-Kemper	Brown University
Lisa Goddard	Columbia University/International Research Institute for Climate and Society
Arun Kumar	NOAA National Centers for Environmental Prediction
Jay McCreary	University of Hawaii
Dimitris Menemenlis	California Institute of Technology/NASA Jet Propulsion Laboratory
Janet Sprintall	University of California, San Diego/Scripps Institution of Oceanography
Bob Weller	Woods Hole Oceanographic Institution
Robert Wood	University of Washington

The US CLIVAR SSC wishes to acknowledge and thank the following Panel Members and other contributors who drafted and/or reviewed content for the Science Plan:

Matthew Barlow	University of Massachusetts Lowell
Tony Barnston	Columbia University/International Research Institute for Climate and Society
Subra Bulusu	University South Carolina
Antonietta Capotondi	University of Colorado
Donald Chambers	University of South Florida
Judah Cohen	Atmospheric and Environmental Research/Massachusetts Institute of Technology

Meghan Cronin	NOAA Pacific Marine Environmental Laboratory
Simon de Szoeke	Oregon State University
Curtis Deutsch	University of Califormia, Los Angeles
Joshua Xiouhua Fu	University of Hawaii
Gregg Garfin	University of Arizona
Alexander Gershunov	University of California, San Diego/Scripps Institution of Oceanography
Allessandra Giannini	Columbia University/International Research Institute for Climate and Society
Benjamin Giese	Texas A&M
David Gochis	National Center for Atmospheric Research
Michael Gregg	University of Washington
Richard Grotjahn	University of California, Davis
David Halpern	NASA Jet Propulsion Laboratory
Yoo-Geun Ham	NASA Goddard Space Flight Center
Meibing Jin	University of Alaska
Markus Jochum	University of Copenhagen
Terrence Joyce	Woods Hole Oceanographic Institution
Igor Kamenkovich	University of Miami
Jennifer Kay	National Center for Atmospheric Research
Hyeim Kim	Stony Brook University
David Lawrence	National Center for Atmospheric Research
James Ledwell	Woods Hole Oceanographic Institution
Sukyoung Lee	Penn State University
Gad Levy	NorthWest Research Associates
Ron Lindsay	Applied Physics Laboratory, University of Washington
Rick Lumpkin	NOAA Atlantic Oceanographic and Meteorological Laboratory
Jennifer Mays	US CLIVAR Project Office
Art Miller	University of California, San Diego/Scripps Institution of Oceanography
Joel Norris	University of California, San Diego/Scripps Institution of Oceanography
Kathy Pegion	University of Colorado/NOAA Earth System Research Laboratory
Balaji Rajagopalan	University of Colorado
Andrea Ray	NOAA Earth System Research Laboratory
Kelly Redmond	Desert Research Institute
Joellen Russell	University of Arizona
Raymond Schmidt	Woods Hole Oceanographic Institution
Siegfried Schubert	NASA Goddard Space Flight Center
Olga Sergienko	Princeton University
Cristiana Stan	Center for Ocean-Land-Atmosphere Studies
Lou St. Laurent	Woods Hole Oceanographic Institution
Fiamma Straneo	Woods Hole Oceanographic Institution
Aneesh Subramanian	University of California, San Diego/Scripps Institution of Oceanography
Liqiang Sun	North Carolina State University
Gabriel Vecchi	NOAA Geophysical Fluid Dynamics Laboratory
Yan Xue	NOAA National Centers for Environmental Prediction
Xiao-Hai Yan	University of Delaware
Chidong Zhang	University of Miami
Rong Zhang	NOAA Geophysical Fluid Dynamics Laboratory
Xiangdong Zhang	University of Alaska

Appendix B:
Acronyms

AAO — Antarctic Oscillation (Southern Annular Mode)

ACC — Antarctic Circumpolar Current

AMIP — Atmospheric Model Intercomparison Project

AMM — Atlantic Meridional Mode

AMO — Atlantic Multidecadal Oscillation

AMOC — Atlantic Meridional Overturning Circulation

AMSR — Advanced Microwave Scanning Radiometer

AMV — Atlantic Multidecadal Variability

AO — Arctic Oscillation (Northern Annular Mode)

BAMS — *Bulletin of the American Meteorological Society*

CalCOFI — California Cooperative Oceanic
Fisheries Investigations

CAMEO — Comparative Analysis of the Marine
Ecosystem Organization

CCSP — Climate Change Science Program

CEOS — Committee on Earth Observation Satellites

CESM — Community Earth System Model

CliC — Climate and Cryosphere Project

CLIMODE — CLIVAR Mode Water Dynamics Experiment

CLIVAR — Climate Variability and Predictability

CMEP — Climate Model Evaluation Project

CMIP — Coupled Model Intercomparison Project

COSMIC — Constellation Observing System for
Meteorology, Ionosphere, and Climate

CPC — Climate Prediction Center

CPRs — Continuous Plankton Recorders

CPT — Climate Process Team

CRED — Center for Research on Environmental Decisions

CTB — Climate Test Bed

DAACs — Distributed Active Archive Centers

DPWG — Decadal Predictability Working Group

DoE — Department of Energy

DOI — Department of the Interior

DIG — Drought Information Group

DIMES — Diapycnal and Isopycnal Mixing Experiment in the
Southern Ocean

DRICOMP — Drought in Coupled Models Project

DWG — Drought Working Group

DYNAMO — Dynamics of the MJO

ECCO — Estimating the Circulation and Climate of the Ocean

ECMWF — European Center for Medium-range
Weather Forecasts

EDV — ENSO Decadal Variability

EDW — Eighteen-Degree Water

ENSO — El Niño-Southern Oscillation

EOL — Earth Observing Laboratory

EPIC — Eastern Pacific Investigations of Climate

ERSST — Extended Reconstructed Sea Surface Temperature

ESGF — Earth System Grid Federation

ETCCDI — Expert Team on Climate Change Detection and Indices

GCM — Global Climate Model

GCOS — Global Climate Observing System

GDIS — Global Drought Information System

GEWEX — Global Energy and Water Exchanges Project

GFDL — Geophysical Fluid Dynamics Laboratory

GHG — Greenhouse Gases

GISS — Goddard Institute for Space Studies

GLOBEC — Global Ecosystem Dynamics Program

GODAE — Global Ocean Data Assimilation Experiment

GOOS — Global Ocean Observing System

GO-SHIP — Global Ocean Ship-Based Hydrographic Investigations Program

GRACE — Gravity Recovery and Climate Experiment

GSFC — Goddard Space Flight Center

GSOP — Global Synthesis and Observations Panel

HadISST — Hadley Centre Sea Ice and Sea Surface Temperature

HYCOM — Hybrid Coordinate Ocean Model

IAG — Inter-Agency Group

IASCLiP — Intra-Americas Study of Climate Processes

ICED — Integrated Climate and Ecosystem Dynamics

ICESat — Ice, Cloud, and land Elevation Satellite laser altimeter

IEAS — Integrated Earth System Analysis

IGAC — International Global Atmospheric Chemistry

IMBER — Integrated Marine Biogeochemistry and Ecosystem Research Project

IOC of UNESCO — Intergovernmental Oceanographic Commission

IOD — Indian Ocean Dipole

ICSU — International Council for Science

IPCC — Intergovernmental Panel on Climate Change

IPCC SREX — Special Report Managing the Risks of Extreme Events and Disasters to Advance Climate Change Adaptation

IPO — Interdecadal Pacific Oscillation

IRI — International Research Institute for Climate and Society

ITCZ — Intertropical Convergence Zone

JAMSTEC — Japan Agency for Marine Earth-Science and Technology

JCOMM — Joint Technical Commission for Oceanography and Marine Meteorology

JGOFS — Joint Global Ocean Flux Study

KEO—Kuroshio Extension Observatory

KESS—Kuroshio Extension System Study

MESA—Monsoon Experiment for South America

MIT—Massachusetts Institute of Technology

MJO—Madden Julian Oscillation

MOCHA—Meridional Overturning Circulation sand Heatflux Array

NAME—North American Monsoon Experiment

NAO—North Atlantic Oscillation

NASA—National Aeronautics and Space Administration

NCA—National Climate Assessment

NCAR—National Center for Atmospheric Research

NCDC—National Climatic Data Center

NCEP—National Centers for Environmental Prediction

NCO—National Coordination Office

NetCDF—Network Common Data Form

NGDC—National Geophysical Data Center

NOAA—National Oceanic and Atmospheric Administration

NODC—National Ocean Data Center

NPGO—North Pacific Gyre Oscillation

NPRB—North Pacific Research Board

NRC—National Research Council

NSF—National Science Foundation

NWS—National Weather Service

OCB—Ocean Carbon and Biogeochemistry Program

OCMIP—The Ocean-Carbon Cycle Model Intercomparison Project

OISST—Optimum Interpolation Sea Surface Temperature

ONR—Office of Naval Research

OOI—Ocean Observatories Initiative

OOPC—Ocean Observations Panel for Climate

OSNAP—Overturning in the Subpolar North Atlantic Program

OSSEs—Observing System Simulation Experiments

PACE—Postdocs Applying Climate Expertise

PCMDI—Program for Climate Model Diagnosis and Intercomparison

PDO—Pacific Decadal Oscillation

PDV—Pacific Decadal Variability

PIRATA—Prediction Research Array in the Tropical Atlantic

PNA—Pacific/North American Teleconnection Pattern

POS—Phenomena, Observations and Synthesis (US CLIVAR Panel)

PPAI—Predictability, Prediction and Applications Interface (US CLIVAR Panel)

PSMI—Process Studies Model Improvement (US CLIVAR Panel)

QuikSCAT—Quick Scatterometer

R2O—Research to Operations

RAPID—Rapid Climate Change

RAMA—Research Moored Array for African-Asian-Australian

Monsoon Analysis and Prediction

REx—Regional Experiment

RISA—Regional Integrated Sciences and Assessments

SALLJEX—South American Low Level Jet Experiment

SAMOC—South Atlantic Meridional Overturning Circulation

SCAR—Scientific Committee on Antarctic Research

SCOAR—Scientific Committee for Oceanographic Aircraft Research

SEARCH—Study of Environmental Arctic Change

SIBER—Sustained Indian Ocean Biogeochemistry and Ecosystem Research

SMOS—Soil Moisture and Ocean Salinity

SODA—Simple Ocean Data Assimilation

SOLAS—Surface Ocean-Lower Atmosphere Study

SOOS—Southern Ocean Observing System

SPARC—Stratospheric Processes and their Role in Climate

SPURS—Salinity Processes in the Upper Ocean Regional Study

SSIWG—Salinity Sea Ice Working Group

SSM/I—Special Sensor Microwave Imager

SSC—Scientific Steering Committee

SST—Sea Surface Temperature

SWOT—Surface Water and Ocean Topography

TAMM—Tropical Atlantic Meridional Mode

TAO/TRITON—Tropical Atmosphere Ocean Array/ TRIangle Trans-Ocean Buoy Network

TAV—Tropical Atlantic Variability

THORPEX—The Observing System Research and Predictability Experiment

TOGA—Tropical Ocean Global Atmosphere Project

UNOLS—University-National Oceanographic Laboratory System

US CLIVAR—United States Climate Variability and Predictability Program

USDA—United Stated Department of Agriculture

USGCRP—United States Global Change Research Program

USGS—United States Geological Survey

VAMOS—Variability of the American Monsoon Systems

VOCALS—VAMOS Ocean-Cloud-Atmosphere-Land Study

WBC—Western Boundary Current

WCRP—World Climate Research Program

WG—Working Group

WMO—World Meteorological Organization

WOCE—World Ocean Circulation Experiment

WWRP—World Weather Research Program

XBT—Expendable Bathythermography

YOTC—Year of Tropical Convection